FERGUSON
A FARMING REVOLUTION

TIM BOLTON

BANOVALLUM
BOOKS

Published in Great Britain in 2021
by Banovallum Books
an imprint of Mortons Books Ltd.
Media Centre
Morton Way
Horncastle LN9 6JR
www.mortonsbooks.co.uk

Copyright © Banovallum Books, 2021

All rights reserved. No part of this publication may be reproduced or transmitted in any form or by any means, electronic or mechanical including photocopying, recording, or any information storage retrieval system without prior permission in writing from the publisher.

ISBN 978 1 911658 22 1

Typeset by Aura Technology and Software Services, India.

10 9 8 7 6 5 4 3 2 1

Contents

Introduction	6
Chapter 1 – The Early Years: Motorcycles, Cars and Aircraft	7
Chapter 2 – The Government Calls	11
Chapter 3 – Ploughs and Tractors	12
Chapter 4 – From Black Days to the Black Tractor	17
Chapter 5 – Into Production: Ferguson Model A	20
Chapter 6 – Another Beginning: Ford and Ferguson	27
Chapter 7 – Difficult Times	34
Chapter 8 – Ferguson Tractor Building in England	36
Chapter 9 – Tractor England	39
Chapter 10 – Ferguson Tractors Overseas	44
Chapter 11 – Ferguson Conversions	47
Chapter 12 – To Diesel with Reluctance	52
Chapter 13 – The Ferguson System	55
Chapter 14 – Other Implements and Accessories	86
Chapter 15 – Bigger is Better?	91
Chapter 16 – Merger	94
Chapter 17 – Across the Ice	96
Chapter 18 – Ferguson in Miniature	101
Chapter 19 – The Legacy	105
Bibliography	113
Index	114

Introduction

Many are familiar with the 'little grey Fergie', even though they may not have any farming connections or ever even driven a tractor. The Ferguson TE20 and its derivatives have become icons of British products, alongside the Morris Minor and Mini, the Land Rover and early Range Rover. Remarkably, its inventor didn't set out to design a tractor and wasn't even a farmer when he did. To say that it arrived by accident is, however, a slight misnomer, even though a series of significant events brought about its introduction – it was the dogged determination of Harry Ferguson that led to its birth. Despite a long gestation and many setbacks, Harry totally believed in his vision – one that succeeded in revolutionising agriculture. The result of his endeavours has not been surpassed nor superseded and is the standard in use throughout the world. ●

Chapter 1
The Early Years: Motorcycles, Cars and Aircraft

Henry George Ferguson (always known as Harry) was born on November 4, 1884, the fourth child of 11 born to James and Mary Ferguson of Lake House, Growell, a small village near Dromore in County Down, Northern Ireland. It is said that Harry had a hard childhood, but whether it was harder than that of other children at the time is debatable. It certainly wasn't that of a privileged child, although the family owned its farm of around 100 acres.

James Ferguson was deeply religious and a hard taskmaster. As with many farming families and indeed many small businesses, the children were expected to help with the day-to-day activities. Harry had no affinity with horses, which provided the power for the heaviest of tasks on most farms, and being of slight build, he was not suited to the heavy manual work of farming.

Two things – the religious views of his father and the heavy farm work – led Harry to give serious thought to emigrating to the USA or Canada, as many Irishmen had at that time. In 1901 Harry's elder brother, Joe, set up a car repair workshop in the Shankhill Road, Belfast, with financial help from their father. In 1902, when Harry was at an advanced stage of planning to leave Ireland, Joe offered him an apprenticeship at the workshop. He accepted it eagerly, and Joe and Harry set off to Belfast with little delay.

While Harry was enthusiastic about motorcycles and cars, he lacked formal training in the subject and attended Belfast Technical College for several years to remedy this. Joe's business, J B Ferguson & Co, gained a reputation as being among the best garages in Belfast and one customer was Thomas MacGregor Greer, who lived at Tullylagen Manor, between Cookstown and Dungannon. This wealthy man owned a series of cars which were often looked after by Harry, who would become close to the family – a relationship which would later prove to be significant.

Around 1904, Harry realised that motor sport could be a very effective way of promoting a business, and both he and Joe took part in a number of events in Ireland. Joe lost interest after an accident involving a local cyclist, while Harry won several trials and races. Four years on, in 1908, Harry began to take a serious interest in aviation and attended air displays at Rheims and Blackpool, where he took measurements and made notes on the construction of the aircraft present. Back in Belfast, Harry convinced his brother that building and flying an aircraft would be good for the business, so during 1909 construction took place. The airplane measured 26ft long, had a 32ft wingspan and was powered by a JAP (J A Prestwich) eight-cylinder engine delivering 35hp, the initial Green engine having been rejected.

The first flight was attempted in Hillsborough Park. The aircraft was towed by car with the wings detached and the rear of the fuselage in the boot. But only short hops were achieved,

1. THE EARLY YEARS: MOTORCYCLES, CARS AND AIRCRAFT

ABOVE: Harry Ferguson built and flew his own aircraft before becoming interested in mechanised agriculture. *National Museums Northern Ireland.*

and bad weather delayed further attempts at flying. After trying several propellers and making several adjustments, a week later Harry tried again, even though conditions were far from ideal. Finally, on December 31, 1909, Harry Ferguson was ready to go.

A reporter from the Belfast Telegraph described the scene: 'The roar of the eight cylinders was like the sound of a Gatling gun in action. The machine was set against the wind, and all force being developed the splendid pull of the new propeller swept the big aeroplane along as Mr. Ferguson advanced the lever. Presently, at the movement of the pedal, the aeroplane rose into the air at a height from nine to 12 feet, amidst the heavy cheers of the onlookers. The poise of the machine was perfect, and Mr. Ferguson made a splendid flight of 130 yards. Although fierce gusts of wind made the machine wobble a little, twice the navigator steadied her by bringing her head to wind. Then he brought the machine down to earth safely after having accomplished probably the most successful initial flight that has ever been attempted upon an aeroplane.'

Harry Ferguson thus became the first person to fly in Ireland and the first Briton to build and fly his own aircraft; feats that on their own would warrant a place in the history books. Flights took place at a number of locations in Northern Ireland, some as long as 2½ miles with the craft reaching heights of 40ft, in preparation for the first aviation event in Ireland, to be held at Newcastle, south of Belfast, on July 23, 1910. The organisers offered a prize of £100 for the first person to make a three-mile flight at the event, billed as Grand Aerial Display and Sports Meeting. Although large numbers of people attended, weather conditions prevented a successful flight and many of the attempts ended with crashes. The town of Newcastle left the offer of the £100 prize open for a month and on August 8, Harry flew for three miles at heights varying from 50 to 150ft, at Dundrum Bay north of Newcastle, to claim the money.

From 1910 the Royal Aero Club had been responsible for issuing Aviator Certificates, which allowed a pilot to take part in displays or contests organised by the club. Certificates were awarded to pilots who successfully completed two distance flights of at least three miles over a closed circuit, one of which had to be completed at a height of at least 164ft. The circuits had to be in a figure of eight around two posts no more than 547 yards apart, and landing within a predetermined distance with a dead engine was also part of the test. Although he produced eight variants of his aeroplane, for whatever reason, Harry never took the aviator test and didn't fly after 1911. The flying exploits led to friction between Harry and Joe, who thought too much time was being taken up and money being spent on the pursuit. He also disliked the aircraft taking up space in the garage.

Encouraged and partly financed by Thomas MacGregor Greer, Harry decided to set up his own garage, initially calling it May Street Motors because Joe objected to him using the family name. It was later renamed Harry Ferguson Ltd. He obtained agencies to sell a number of makes of car including Vauxhall, which at that time was an upmarket sporting make.

Harry again took part in a number of events in which he was successful and later, at the beginning of the First World War, he began to sell tractors, obtaining an agency to sell the Overtime tractor (called the Waterloo Boy in the US). To promote the tractor, he arranged a series of ploughing demonstrations which didn't always turn out as expected, as many farmers were reluctant to change from horsepower and took every opportunity to deride Ferguson's efforts. Often the results of using a three-furrow Cockshutt plough behind the tractor were not as neat as a single furrow horse plough and it was said that pursuit of the perfect finish was what led to the reluctance by farmers to accept the tractor over the horse.

Even so, these ploughing demonstrations would stand Ferguson in good stead and, unknown to anyone at the time, prove to be the beginning of one of the greatest advances ever made in agriculture. ●

ABOVE: Demonstrations were used extensively to highlight the benefits of linked ploughs with depth control.

1. THE EARLY YEARS: MOTORCYCLES, CARS AND AIRCRAFT

ABOVE: The first ploughs were attached to a Model T Ford with Eros tractor conversion. *National Museums Northern Ireland.*

ABOVE: This early Ferguson plough attached to a Fordson F precedes the Duplex plough. *National Museums Northern Ireland.*

Chapter 2
The Government Calls

Many working men lost their lives in the First World War and food production suffered as a result, so the British Government asked for vast acreages of pastureland to be turned over to food production. The war also resulted in the loss of farm horses, a situation thankfully remedied by Henry Ford building and shipping 5,000 tractors for the Ministry of Munitions.

Despite the criticism from some farmers, Harry Ferguson and his employee Willie Sands gained a reputation as tractor ploughmen, so the Irish Board of Agriculture asked Harry to improve the efficiency of Ireland's tractors during 1917 to enable more food to be produced.

Ferguson and Sands toured Ireland in a large car provided by the Government, which had been modified so that they could sleep in it if necessary. Having adjusted many tractors and ploughs during the tour, the pair realised there were basic problems associated with tractor ploughing. The tractors were heavy and often unwieldy, most having been built for the US market and imported into Britain. The lightest, the Overtime, weighed more than two tonnes, and many weighed more than four tonnes.

Ploughs were also considered to be too heavy in their construction and little altered in design from that of horse ploughs. The biggest problem identified was the method of connecting the plough to the tractor, which again had been carried over from the age of ploughing with horses. When a horse-drawn plough hit an obstacle, such as a rock or tree root, very often the horse or horses would be forced to stop, not having the power to carry on regardless. The ploughman would lift the plough clear of the obstacle and carry on, or occasionally he would spot the obstruction and lift the plough in advance. With tractors having more power and the driver on the tractor rather than behind the plough, a new problem occurred. When the plough hit a root or stone, either the plough was damaged or the force of the plough against the obstruction caused the front of the tractor to rear up, sometimes flipping over with disastrous consequences.

The tractor manufacturers' answer to the problem was to apply more weight to otherwise heavy tractors or extend the length of the tractor to prevent this rearing up; neither solution was satisfactory. There was also the problem of manoeuvrability: a long heavy tractor and a plough behind it (trailer or trailing plough) made turning at the end of each furrow difficult. Ferguson realised the present ploughing set-ups didn't make use of the forces generated while ploughing and resolved to design a plough that overcame the problems. ●

Chapter 3
Ploughs and Tractors

His dislike of heavy tractors led Harry to choose the Ford Model T Eros tractor conversion to draw his plough. The very popular Model T spawned many conversions, including some that turned the car into a simple tractor and one of these was the Eros.

With much of the car body stripped away, two small spur gears were attached to the ends of the axle where the wheels would have fitted. Two much larger, tractor-type wheels sitting on their own axle and sub-frame were attached to the car chassis and driven by the spur gears engaging with internal toothed gears fitted to the inside of the larger wheels' rims. This arrangement lowered the gearing and speed of the Model T considerably to that suited to farm work.

Harry designed a two-furrow plough and Willie Sands built it. The plough, which was only about a third of the weight of a conventional one, was attached underneath the back of the Eros and forward of the rear axle to prevent the rearing up associated with trailing ploughs and have the effect of forcing the wheels into contact with the ground. The weight of the plough was counterbalanced by springs, enabling the driver to raise and lower it in and out of the ground by means of a lever at the side of the seat.

One of the first demonstrations of the new plough took place at Coleraine. The prototype for demonstrations had beams of cast iron, whereas it was intended that on production versions the beams would be made of cast steel, a much stronger and more suitable

ABOVE: A tractor conversion of a Model T Ford car, similar to that used by Harry Ferguson.

material for the task. However, onlookers were unaware the plough they were observing was a prototype, so were unimpressed when a rock was struck, and it was shattered.

During 1917 Henry Ford introduced a new tractor. He was another farmer's son and had for several years built experimental farm tractors with a view to improving farm efficiency and, of course, having another product to sell. The new tractor was built on a unitary system. The castings of the engine, radiator, gearbox and rear axle, were load bearing, rather than being fitted into a frame that bore the loads. While this wasn't new as regards tractor design, it was a first for a mass-produced tractor and enabled the vehicle, known as the Fordson F (the name Fordson was used as this was a separate venture from the Ford car side of the business), to weigh less than 25 cwt, which was a lot less than contemporary tractors. In agriculture a heavy tractor tends to compact the soil, affecting drainage and productivity.

ABOVE: The Ferguson duplex plough was designed for the Fordson F.

In the spring of 1917, the British Board of Agriculture asked the Royal Agricultural Society of England to carry out tests on two imported Fordson tractors. The tests took place in Cheshire on April 24 and 25 and showed the tractor was suitable for use on British farms.

The Government wanted the tractor to be produced in Britain and drawings were sent over from Ford. After careful costing however it was realised the tractor couldn't be produced economically. Eventually a factory was built in Cork, Ireland, and while this was being carried out and as the situation was desperate with ships carrying food being sunk on a regular basis, 5000 tractors were imported from Detroit.

Henry Ford brought forward the production of the new tractor by adding an extension to an existing factory, the first tractors arriving in early December 1917 and on sale in Britain for £250, much lower than a Model T with Eros conversion.

When Henry Ford's representative Charles Sorenson arrived to discuss building the tractor in Britain, Harry and Willie Sands went to see him in London. Here Ferguson pointed out that the Fordson was little better than a team of horses, in that it still pulled trailed implements, and that integrated implements were the way forward to achieve better efficiency in agriculture. Sorensen agreed with him, but they were so busy designing and developing the tractor that there was no time to consider a linked system of implements and such a system would be difficult to achieve anyway.

Sorensen was impressed with Ferguson's drawings and encouraged him to continue with plough development, though this caused some problems for Ferguson as he had stocks of ploughs designed for the Eros, but felt that a different style of plough would be better suited to the Fordson. It also led to some conflict with fellow directors of Harry Ferguson Ltd, who thought he was spending too much time and money chasing his agricultural goals. Macgregor Greer was one who supported him.

3. PLOUGHS AND TRACTORS

Harry designed a plough to fit the new Fordson tractor, based on the same principles as the earlier plough in that it fitted directly to the tractor, with similar basic geometry to prevent rearing up when an obstruction was encountered. Two brackets attached the plough to the Fordson, a lower one in place of and in the style of the drawbar and one higher up that was attached to plates bolted to the axle. Because of these two fixings the plough became known as the Duplex Hitch.

The weight of the plough was counterbalanced by a long strong spring and raised or lowered to the ground by a lever operated by the driver. As with the plough for the Eros, there was no depth wheel to limit how far the ploughshares sat in the ground, the geometry supposedly holding it correctly. Being directly attached to the tractor though, under certain undulating ground conditions a constant depth of furrow could not be maintained. This problem was to occupy Harry and his team for a long time, but when solved was the making of Ferguson.

Without much of the apparatus of the conventional trailed plough, the Duplex Hitch weighed considerably less. The layout was such that Harry was able to obtain a patent for it – one of a large number he would be granted over the years. As the plough was designed to fit the Fordson tractor, Harry decided he should show it to Henry Ford. Having already been in contact with Charles Sorensen, he wrote to him to ask about demonstrating the plough and the reply was favourable.

Harry took Willie Sands with him on the trip to see Henry Ford. They crossed the Atlantic by ship to Quebec and then by train to Windsor, just over the Canadian border from Detroit. The Ford headquarters at Dearborn lay just a few miles to the south west of Detroit and the vast Rouge River plant carried out most of the processes required for car manufacture.

The demonstration plough they took with them had beams of bronze as they did not have access to facilities for casting steel. At the Ford plant they were able to get the beams cast in steel; one of the requirements for a strong, light plough.

A Fordson tractor was provided for the demonstration, at which both Henry Ford and Charles Sorensen were present, and it was clear that Ford was impressed. Misunderstanding Harry's character and capabilities, Ford, through Sorensen, made him a job offer, which was politely declined even though the proposed salary was increased several times. Undeterred, Ford offered to buy the patent for the plough but again this was declined. Nothing concrete came of the meeting, other than Henry Ford encouraging Harry to continue his work on the plough and keep in touch.

Once again, Harry's fellow directors were less than impressed. They saw the trip as a waste of time and money and tried to persuade Harry to abandon his work with ploughs. With the future uncertain, Willie Sands decided to leave and set up a motor repair workshop with his father in Belfast. Archie Greer, who had worked for Harry for a number of years, took on the role of his main helper.

Fixing the plough directly to the rear of the Fordson solved many of the problems associated with trailing ploughs, except for that of maintaining accurate depth. A depth wheel running on the unploughed land would have solved the problem but would erase some of the benefits associated with the linkage plough, such as preventing rearing up. After much experimentation with extra springs and linkages, Harry concluded the only way to solve the depth control problem was to fit a depth wheel to the plough, even though this would negate weight transfer to the tractor. Harry reasoned this solution would be acceptable as no other plough on the market provided weight transfer and because the features of the Duplex plough would be enough to give it a significant advantage over other ploughs.

When Harry and Willie Sands visited America in 1920, they were amazed at the size of the fields, which required a number of tractors working in unison to cultivate them. With this in mind, Harry felt America offered great potential for the sale of ploughs and in 1922 again made the trip, this time accompanied by his friend John Williams.

The plough was shown to John Skunk, who operated a large metal-working business in Bucyrus, Ohio, and seeing the potential he agreed to manufacture it. The local Chamber of Commerce held a banquet at the end of May 1922 to celebrate the signing of the agreement. Perhaps being over optimistic, Harry announced 1000 ploughs would be produced by the autumn and the first full year's production would consist of 50,000 units. Soon he went on to announce the whole of the Skunk works would be needed and every plough manufacturer would eventually adopt his principles.

For whatever reason (it was never clear, although a few theories were put forward) John Skunk didn't manufacture a single Ferguson plough. Harry and John Williams returned to Belfast and gave an interview to the Press without realising the plough would not be manufactured by Skunk. The news forced Harry to return to America and in Mansfield, Ohio, he held discussions with Roderick Lean, an implement manufacturer who, among other products, made a set of disc harrows specifically for the Fordson tractor. An agreement was signed for Lean to manufacture the Ferguson plough.

Problems occurred after Harry returned to Britain. There were times when ploughing in difficult conditions that the lack of weight transfer would cause the tractor wheels to lose grip and spin, much as he envisaged. The problem of depth control without a wheel running on unploughed ground also still needed to be resolved.

Harry visited Willie Sands at his Belfast workshop. His former colleague was becoming disillusioned with repair work as the rewards were not what he expected. After a long conversation it appeared Sands had enjoyed his time working with Harry on development, so when he was made an offer to go back and work on the depth control problem, he accepted. As manufacture of ploughs by Lean had already started, not only would the depth control problem have to be solved but solved in such a way that new parts could be retro-fitted to existing ploughs.

A patent was applied for in 1923 that described a small wheel fixed to the rear of the plough and riding in the bottom of the furrow. In reality this was replaced by a hinged steel plate that could be adjusted by a threaded lever situated at the front of the plough. This arrangement was cheaper to produce than fitting a small wheel.

With one problem solved, another came along. In 1924 the Lean company got into financial difficulties and went bankrupt. Once again, the search was on for a manufacturer of Ferguson's plough and in 1925 Harry and John Williams travelled to America to find one. This time, instead of seeking out an existing manufacturer he formed Ferguson Sherman Incorporated with brothers Eber and George Sherman, whom they had met on previous visits to America, and a factory was set up in Evansville.

There were excellent advantages in joining with the Sherman brothers. They were the main Fordson tractor distributors for New York State and Eber had been Ford's export manager. Harry, his wife Maureen and John Williams spent several months working on the business and the large quantities of ploughs eventually produced sold well.

Back in Northern Ireland and together with Willie Sands and Archie Greer, Harry investigated linking other implements in unit with the tractor. While the problem of depth control for a plough had been solved with the plate running in the bottom of the furrow, the same principle could not be applied to cultivators or ridgers. When an implement such as a plough or cultivator is engaged in the ground, a resistance force is encountered, and the team realised this force could be utilised to provide depth control. In 1925 a patent was applied for concerning depth control and granted the following year. Various ways of obtaining depth control were tried,

3. PLOUGHS AND TRACTORS

mostly mechanical, using pulleys, cables, levers and clutches, although electric motors were also proposed, most significantly a hydraulic system.

Both the mechanical and hydraulic systems could be designed to fit the Fordson tractor and it was thought the mechanical system would be the easiest to develop, but it didn't prove so; the system looked ungainly and the cone clutches wore very quickly.

The use of a hydraulic system to raise and lower implements was not a new idea, but neither was it in production – no one had linked hydraulic lifting and lowering of a plough with depth control. Some work had already been carried out by Sands and Greer on a hydraulic lift that could be fitted to the Fordson tractor, with the arrangement relatively simple. A pump supplied oil under pressure to a hydraulic ram, which was connected to lift arms, so that when the pump fed oil to the ram, the pressure extended the ram and raised the arms. Lowering of the arms was achieved by releasing the hydraulic pressure through a valve to allow gravity to force oil out of the ram.

As this system was not in production, Harry and the team had to design and build suitable components, not only for raising and lowering but also one that was sensitive to the draft of the implement to allow small amounts of movement to adjust the draft. A single lower arm central to the two upper lift arms was tensioned when the resistance of the plough acted upon it and this tension was used to alter the draft of the plough. A progression of hydraulic pumps was designed and built – some complex, others simpler – with an early variable stroke pump abandoned because it couldn't be controlled to the accuracy required.

Constant delivery pumps that dumped hydraulic oil back into a reservoir when pressure was not required were tried, but all versions were eventually discarded due to vibration problems. Testing was carried out at a farm near to Belfast, using a hydraulic linkage fitted to a Fordson tractor and although there was no eureka moment, constant testing and adjustment steadily improved progress. •

ABOVE: While working for the Irish government, Harry Ferguson discovered that tractors were too heavy and used outdated ploughs.

Chapter 4
From Black Days to the Black Tractor

The fortunes of tractor manufacturers in America underwent dramatic change during the 1920s. At the beginning of the decade Ford, with its Fordson tractor, achieved 75 per cent of tractor sales, while the other manufacturers fought back with mergers, entirely new models and customer incentives. The result was that in 1928, with a worsening economy, Ford, along with General Motors, decided to pull out of tractor production in America. The repercussion of this momentous decision was the ruination of the Ferguson Sherman plough business, as the ploughs were designed specifically for the Fordson tractor.

Some 2000 ploughs were still unsold, but Ford decided to resume production of tractors at the factory in Cork and import them into America, so they proved not too difficult to get rid of. But it wasn't enough to save the plough business.

Work continued on solving the depth control problem but with lower funding, and Harry told Willie Sands and Archie Greer he would most likely have to cut their wages. Sands decided to leave for a second time and set up a bus service using a Bean omnibus.

Harry kept faith in the principle of draft control and felt it would be a major breakthrough when fully developed, but again he had to convince his fellow directors to allow him to continue researching and developing it. He reached a stage whereby the hydraulic pump worked continuously and when hydraulic pressure was not required, a valve opened and hydraulic oil short circuited the ram and went back into the oil sump. While this was a better arrangement, it could cause heating and aeration of the oil, leading to inconsistency of the system.

Harry realised another problem with trailed implements was apparent when the tractor turned either to the left or right, and the single pin attachment caused the implement to turn on a tighter radius. In crops, and particularly those planted in rows, this would cause the implement to run over and destroy plants. The problem could be solved fairly easily by mounting the implement forward of the tractor's rear axle and many manufacturers overcame this effect by mid-mounting implements. For Ferguson, though, this was not the answer; he wanted depth control to be applied to all ground-engaging implements, not just ploughs, and wanted all implements attached to the tractor in the same way.

The answer was relatively simple. As with solving the rearing up problem, it was a matter of altering the geometry of the linkage. The two upper links were angled inwards towards the tractor, so their virtual converging point was in the centre of the tractor, which gave the same effect as having a mid-mounted implement, and in 1928 Ferguson applied for a patent covering this.

It was felt the hydraulic system fitted to the Fordson was now satisfactory and several demonstrations were carried out in Ireland in the hope that a manufacturer would be interested enough to start production of the system. Later in 1928, Harry, accompanied by his wife, went to America to contact companies that might be interested in producing the system, and although Allis Chalmers took out an option on it, this interest went no further.

4. FROM BLACK DAYS TO THE BLACK TRACTOR

Several British companies showed interest in the system but for a number of reasons – unsuitable tractor and lack of finance being among them – these companies did not follow up the opportunity. Not only did Harry continue to demonstrate the system, he also gave speeches extolling its advantages, but depth control was not perfect and some were unable to see the benefits, as under certain conditions jerkiness in the system caused unevenness in the bottom of the furrows. Harry was undeterred and firmly believed in the merits of depth control and the ability of his team to solve the outstanding problems.

Although the demonstrations stirred interest, a manufacturer was not found and Harry took the tractor to England in the hope of finding a company to take on at least the manufacture of the system, but preferably of a tractor incorporating the system.

At a demonstration held at Ascot, the Prince of Wales and the Duke of Kent attended and were said to be impressed. Later a representative of Morris Motors was present and reported back to William Morris, who was interested enough to offer to be involved in the design and manufacture of a tractor and range of implements based on Ferguson's work. Although an agreement was drawn up and approved, it was withdrawn before it could be signed, the reason unknown. It could have been that in some quarters it was felt that manufacturing anything farm-related was unrewarding, but also that Morris was unlikely to have suitable products already in production that, with modification, could be incorporated into a new tractor.

Such was the nature of Harry that although he was disappointed by the numerous setbacks, he refused to deviate from his belief in his inventions and foresight. Despite his demonstrations and speeches extolling the virtues of linked implements and draft control, he had not been able to get any company to agree to manufacture even the hydraulics and linkage, let alone a new tractor.

Although the Fordson tractor had been used for experimental work and demonstrations, it wasn't, in Ferguson's estimation, ideally suited to be the basis of what he envisaged. At under 25 cwt the Fordson was a comparatively light tractor, with many of its contemporaries weighing up to three tonnes. However, Harry considered the Fordson to be too heavy; with his plough linkage providing weight transfer, a heavy tractor was unnecessary.

In 1931, realising that no company was going to build an entirely new tractor and that none of those in current production were ideal, he decided the only way forward was to design and build a prototype tractor incorporating his ideas.

The team of Harry, Sands (who had now returned, having sold his bus business) and Greer, set about laying out ideas for a lightweight tractor incorporating the hydraulic draft control system. During 1933, Greer and draughtsman John Chambers produced detailed drawings for the manufacture of the Ferguson tractor, which was based on the unit principle favoured by Henry Ford.

Gear manufacturer David Brown of Huddersfield was chosen to make the transmission and steering components. A Hercules four-cylinder engine was imported from America. This side-valve engine produced 18hp and use of aluminium alloy castings resulted in a tractor that weighed only 16 cwt. The tractor was assembled at the May Street garage and painted black, thereafter becoming known as the 'Black Tractor'.

To give no doubt as to its origin, the name 'Ferguson' in script was cast into the front of the radiator header tank. Steel wheels were fitted; those at the rear having spade lugs bolted to the outside of the rim to aid traction in muddy conditions, and the front ones having a thin vertical steel band around the circumference to aid steering in wet and muddy conditions. To further aid steering, particularly on tight turns, the two rear brakes could be operated independently.

The hydraulic linkage was modified. Instead of two upper arms and one lower one, this was reversed to give one upper and two lower and the lower arms still in tension to provide input to the hydraulic depth control. After much testing, it was found that using the single top link in compression provided much better depth control input. The problem of the hydraulic oil getting too hot, leading to aeration and inconsistency in the depth control, was finally solved. Instead of oil being bled off from the output side of the hydraulic pump, the valve was put on the input side, cutting the oil supply to the pump when not needed. Alongside the new tractor, complementary implements were produced – besides the plough there was a ridger or bouter for potato planting and ridging up.

Much of the testing was carried out on fields surrounding Tullylagen Manor, with the permission of Thomas MacGregor Greer. Being patriotic, Harry would have preferred his inventions to be taken up by a manufacturer in Ulster, but early on it was apparent this would not happen. ●

Chapter 5
Into Production: Ferguson Model A

Ferguson brought the 'Black Tractor' to England, again giving demonstrations with the hope of finding a company interested enough to produce it. One such company was the Craven Wagon and Carriage Works, a subsidiary of a large steel company. Also interested was David Brown, the grandson of the founder of the company by the same name. Craven Wagon went as far as signing an agreement however, such was the nature of Ferguson, that any suggestions of modifications or improvements to his design were rebuffed. He wanted complete control and even though Craven Wagon was part of a large steel company, he insisted that American steel be used. Relations very quickly soured and Craven Wagon was relieved when the young David Brown enthusiastically took over the manufacturing agreement, despite opposition from his father and warnings from senior people at Craven Wagon.

The production tractor was very similar in overall design to the 'Black Tractor' and the casual observer would have had difficulty in distinguishing them, apart from the colour. As could be expected of both Ferguson and a large engineering company, the tractor was constructed to a high standard and even the nuts were specially treated to prevent 'rounding off' of the hexagon corners.

The tractor was designated Ferguson Model A, although in many quarters it was and is referred to as a Ferguson Brown. Instead of the Hercules engine of the prototype tractor, the production version used a four-cylinder side valve engine built by Coventry Climax which produced 20hp, and the first few hundred tractors used this engine. When Coventry Climax was unable to supply more engines, David Brown took over the manufacture of the engine, which was then produced with minor modifications to the original and a new company, Ferguson Brown Limited, was formed.

Harry wanted the production tractors to be the simple black of the prototype but was persuaded to use battleship grey, a colour that would come to define the Ferguson tractor. The Model A was available with steel or pneumatic wheels, the latter making the tractor £40 more expensive, at £270, than the steel-wheel version. Both types of wheel could be

ABOVE: A small plate at the base of the radiator states that this early Ferguson A was built by David Brown Tractors Ltd.

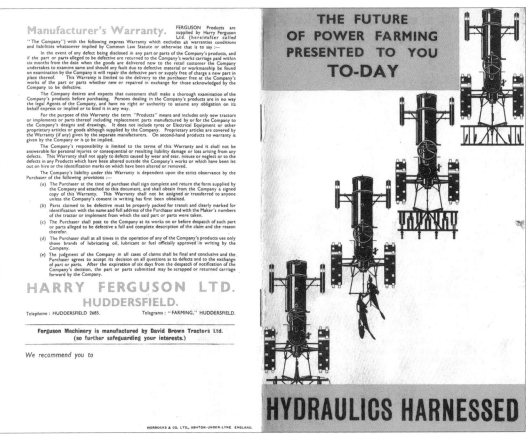

ABOVE: The cover of a brochure setting out the benefits of the Ferguson hydraulic system.

adjusted to give different width tracks for row crop work and for those with steel wheels that needed to use the tractor on public roads, steel bands were available to prevent the spuds or lugs damaging the road surface.

Much was made of the hydraulic depth control, as this had occupied Harry and the team for so many years. Strangely though, an industrial version of the tractor on pneumatic tyres but without any hydraulics was listed as available at £244. A range of four implements to complement the tractor were available: a two-furrow plough; two cultivators, one with spring-loaded tines to prevent damage when an obstacle was struck and one for row crop cultivation whereby the tines could be adjusted to fit the rows; and a three-furrow ridger or bouter. All were priced at £26. For six pounds ten shillings (£6.50), a belt pulley could be fitted to the right-hand side of the transmission housing to enable stationary machinery to be driven by the tractor. For those farmers wishing to spread the cost of the tractor, it could be bought on hire purchase over 12, 18 or 24 months with a down payment of £69.

The Model A weighed just 16½ cwt, around two thirds of the weight of the Fordson. With a turning circle of 21ft, the tractor was very manoeuvrable, particularly as with the implements being in unit it was easy to reverse, something virtually impossible to achieve with trailed implements.

Much was made of the manoeuvrability of the tractor and demonstrations were set up whereby a small plot of land measuring 27ft by 20ft was fenced off with just a single entrance and exit on the long side. The tractor would plough every part of the enclosed ground,

5. INTO PRODUCTION: FERGUSON MODEL A

ABOVE: Ferguson A was originally only available on steel wheels.

ABOVE: Little altered on the control lever design over the years. A page from an early brochure.

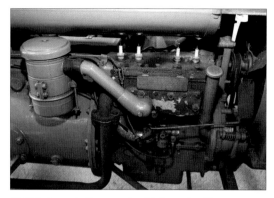

ABOVE: An early Ferguson A with Coventry Climax engine.

ABOVE: Identification plate on the Ferguson A lists a number of patents relating to the tractor.

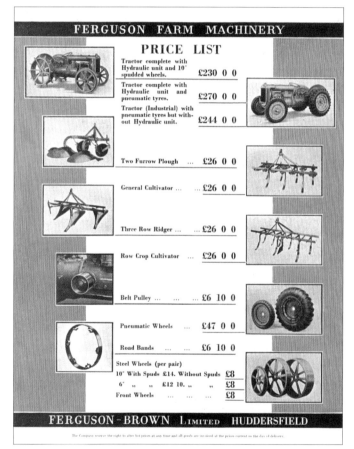

ABOVE: Price list for the Ferguson A.

using the ability to reverse to maximum advantage, and much was made of the small size of the plot and the fact that no wheel marks were left. This type of demonstration was used to very good effect well into the 1950s. There was a drawback though, and the driver of a Model A had to be very aware of it not to get caught out: the hydraulic pump only operated when the tractor was in motion and if the clutch was disengaged the drive to the pump ceased and the implement could not be raised. This was particularly tricky when the tractor was at the end of the furrows and approaching the boundary of the field, when the plough needed to be raised in plenty of time to prevent it being stuck in the down position.

Although Model A and its implements were revolutionary, this did not necessarily convert into instant success. There were two problems: buying the tractor on its own would be pointless as the linked implements were integral to the system, and the minimum outlay for a farmer to acquire a tractor and two implements was £282. At the time the Fordson tractor – now the improved, designated Fordson N and built at Dagenham, Essex – was priced at £145. Also, farmers would already have plenty of implements to cover their needs. Even up to

5. INTO PRODUCTION: FERGUSON MODEL A

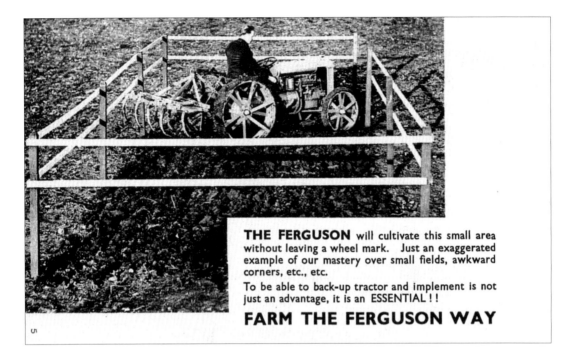

THE FERGUSON will cultivate this small area without leaving a wheel mark. Just an exaggerated example of our mastery over small fields, awkward corners, etc., etc.

To be able to back-up tractor and implement is not just an advantage, it is an ESSENTIAL ! !

FARM THE FERGUSON WAY

ABOVE: First used when demonstrating the Ferguson A, cultivating a small area was a feature of demonstrations well into the 1950s.

LEFT: Ferguson A with Ferguson plough at work in 1936.

the late 1940s, there were companies offering conversions so horse-drawn equipment could be used behind a tractor.

It would only be wealthy, enlightened farmers who bought the Ferguson, and in the mid to late 1930s Ford produced almost 60,000 Model N tractors from its Dagenham plant against a total production run of just 1250 Ferguson A tractors.

The tractor with the number one serial number was purchased by Thomas MacGregor Greer, who specified pneumatic tyres. When rear mudguards were introduced, he had them fitted with extra metal added to form wide flat

ABOVE: A Ferguson A and Fordson N sit alongside each other. Beginning of one era, end of an earlier one.

ABOVE: Rare orchard version of the Ferguson A sits alongside a TE A20 in local authority colours.

5. INTO PRODUCTION: FERGUSON MODEL A

ABOVE: Ferguson A tractors are sought after even in this condition.

tops, providing extra protection and giving this tractor a distinctive and unique look.

Harry wanted to lower the price of the tractor to make it more competitive, with the added benefit of hopefully increasing production volume, while David Brown thought it needed more power, most likely meaning an entirely different engine would be needed. He resisted a price reduction and inevitably, neither agreed with the other and a split between the two began to form. David Brown started to plan a more powerful tractor to his own design, and Harry sought another manufacturer to build tractors incorporating his ideas and patents.

Towards the end of the 1930s, Ferguson Brown Limited ceased to exist and a new company, David Brown Tractors Limited, took over selling the remaining tractors, reduced in price – 'Now Under £200' declared the advertisements – in an effort to clear remaining stocks. ●

Chapter 6
Another Beginning: Ford and Ferguson

Once again Harry looked across the Atlantic for a solution. When he had met Henry Ford some years earlier, they got on well together; despite an age difference, they were from farming families and both realised mechanisation in farming was the way forward to increase productivity and feed the growing population. Also, Henry Ford had built experimental tractors long before he built his first car.

Although the Ferguson Sherman plough business was finished, Harry kept in touch with the Sherman brothers and in early 1938 Eber came to England to see a demonstration of the Ferguson Model A. Eber was also still on good terms with Henry Ford and suggested he should look at Ferguson's tractor. Ford responded that he would like to see the tractor – just the answer Ferguson had hoped for.

Perhaps unknown to Ferguson, Henry Ford had once again been toying with the idea of producing a tractor. During the latter part of the 1930s several prototypes were built and tested, some unusual and others more conventional. Ford considered his newly introduced V8 engine could be used, and a three-wheeled tractor with a huge driving wheel and two smaller stabilising wheels was built and tested using the V8. The tractor resembled one built in the 1920s with that large wheel and the driver sitting forward of it to give a clear view. Another prototype was a little more conventional. An Allis Chalmers tractor had been obtained and stripped down. The Ford prototype was a little larger but built in the same tricycle row crop style with the two front wheels set near together. This tractor used a six-cylinder engine and the grille and bonnet from a Ford truck.

In October 1938 Ferguson and John Williams sailed to New York, a Ferguson tractor and implements already shipped before them. On the voyage Harry encountered Leon Clausen, of the J I Case company, a large US tractor manufacturer, and discussed with him his tractor design ideas. Clausen showed no interest, believing heavy tractors were the answer. They arrived in New York on November 1. They met up with the Sherman brothers, un-crated the tractor and implements and arranged for them to be transported to Dearborn. They had 48 hours to prepare the demonstration, which took place over six days at the nearby Ford family farm. To test the Ferguson tractor and its depth control mechanism to the fullest extent, ridges and dips were artificially introduced into the surface of the field which would pitch the tractor up and down and evaluate the accuracy of the depth control. Measurements of the depth of the furrows were taken and found to be very consistent. Towards the end of the test, on Henry Ford's instructions, the Ferguson tractor was taken to a different field, which had been baked hard, and once again the vehicle and plough handled themselves well, easily ploughing the tough surface.

Ford was obviously impressed, as he had been some 20 years earlier when the duplex plough was demonstrated to him and when he'd made offers to Ferguson to become a highly-paid employee or sell his patents for a large sum, both of which were rejected.

Two chairs and a table were brought out to the field, and Henry Ford and Harry sat opposite each other. In the background was the Ferguson Model A, and on the table an early version of the model that Harry, and much later his sales staff, used to effectively demonstrate the advantages of the three-point linkage.

The clockwork-powered model tractor would pull a trailing plough and a mounted plough in turn along a wooden track, in the centre of which an obstruction had been fixed. When the share of the model trailing plough hit this obstruction, it forced the front of the tractor upwards, but when the linked plough share hit the obstruction the front of the model tractor stayed down.

In the middle of an American field, the two visionaries agreed to work with one another on a new tractor incorporating Ferguson's hydraulic depth control. No formal contract was drawn up, and this historic meeting became known as 'The Handshake Agreement'. Ford would manufacture the tractors and Ferguson would market them. Since many different photographs of the pair sat at the table exist, it could well be that this was stage-managed for publicity purposes and that the real discussions had already taken place. As far as this author is aware, no photograph exists of the pair shaking hands, although it has been said that they stood up from the table and did this.

One would have thought that these two men, well-briefed in business and the ways of the world, would have wanted a written agreement, the lack of which later proved to be a bad decision. Ferguson had taken out many patents on his inventions and one would have thought that these needed to be protected.

One part of the agreement was that the new tractor would eventually be built in England at Ford's Dagenham plant. For reasons explained later, this never happened. Harry and John Williams returned home and the working relationship with David Brown was formally dissolved.

The Ferguson tractor remained in America, where it was dismantled and examined by Ford workers to better understand the workings of the hydraulic depth control. Prototypes were built incorporating Ferguson's hydraulic depth control system, with the resulting two tractors smaller than the Ferguson Model A. One was fitted with a two-cylinder horizontally

LEFT: The famous 'handshake agreement' – Henry Ford sits opposite Harry Ferguson, an early clockwork demonstration model between them.

ABOVE: Ford tractors with Ferguson system used a Ford side-valve engine. TVO vaporiser manifold fitted.

opposed engine and the other with a four-cylinder engine from a British Ford truck.

Harry and his team returned in January 1939 to finalise details for producing the new Ford-built tractor. While away, they had been working on a power take-off shaft for this tractor and an adjustable front axle. Much of American farming involves crops being grown in rows and as the width of these rows varies according to the crop, it was essential the track width of a tractor could be adjusted to suit. The front axle consisted of three elements: a pivoting centre section and two outer sections, left hand and right hand, which carried the king pins and wheel hubs. A series of holes were formed into all three sections so they could be easily bolted together at differing widths.

To overcome the need to adjust the steering geometry, a drop arm was fitted to each side of the steering box instead of just the one fitted to most tractors, and separate drag links went to the steering arm of each wheel. Radius arms on each side prevented fore and aft movement of the axle. This design of axle would be used on Ferguson tractors until the Massey-Ferguson 65 of 1958 and a patent for it was granted in July 1939. The design was later used by other tractor manufacturers, notably Ford on its Dexta.

Although a prototype power take-off shaft was fitted to a Ferguson Model A tractor, it was considered unsuitable for production purposes and would certainly have been much too ungainly for the new tractor. The Model A power take-off shaft had an external drive taken from the belt pulley to a housing on the top of the rear axle and was engaged and disengaged via a lever in front of the belt pulley housing, altogether an unsatisfactory arrangement. The design of the system for the new tractor had a shaft within the transmission casing and terminated centrally beneath the rear axle, an ideal position for driving implements such as mowers.

Henry Ford promised to put huge resources behind the design of the new tractor and was true to his word; a prototype was ready by March 1939. Harry insisted on having control over the design process and delays often occurred

6. ANOTHER BEGINNING: FORD AND FERGUSON

because of his involvement. Much of what he wanted to be incorporated, such as an overhead valve engine and a four-speed gearbox, were not in the final specification and neither were the specially made components. Ford knew how to keep costs down and speed up development by using existing components, particularly for the front hubs and transmission parts. The engine was based on one half of the Ford V8 engine, which was of side valve configuration.

A similar driving position to that used on the Model A was incorporated on the new tractor. The driver sat slightly forward of the rear axle with legs either side of the transmission casing and two simple peg extensions of the radius arm fixings provided footrests. The hydraulic control lever placed on the right-hand side was within easy reach. This type of driving position contrasted greatly with that of the Fordson N tractor, where the driver sat to the rear of the axle and was surrounded with sheet metal that gave some protection.

Either side of the drawbar, perforated steel platforms gave the driver standing space, a useful feature though a little precarious, enabling both relief from sitting on long days and the ability to monitor and adjust trailed implements, mainly ploughs. The prototype tractor was fitted with a simple bonnet, very like that of an Allis Chalmers model B. Because of steel restrictions, the first production bonnets were in aluminium and the elegant grille in chevron-style cast in the same material. The enclosure of tractor working components with sheet metal was a 1930s trend, following in the footsteps of the Art Deco movement.

While the Ferguson Model A was technically advanced, style-wise it was like the Fordson F in that the radiator and fuel tank were regarded as functional items and left exposed. A number of manufacturers, notably John Deere, Allis Chalmers and International Harvester, had updated models by enclosing the front of the tractor in sculpted sheet metal – these would come to be known as 'styled' versions. Fordsons in England did not update the look of its tractors until the new Major of 1951. The new Ford tractor would go further, style-wise, than other tractors: the front grille and bonnet (hood in American parlance) were as one and covered not only the radiator but also the fuel tank and battery, a hinged cover allowing access to the fuel tank. The mudguards in a clamshell shape were economical in their quantity of steel.

Production of the new tractor, designated 9N (9 for the year 1939 and N for the Ford code for tractors) commenced on June 5, 1939. By the end of June only 77 tractors had been completed, the slow pace being deliberate to allow workers to become familiar with the assembly process. The colour of the tractor was the familiar Battleship Grey as used on the Model A.

Ferguson wanted the tractor to be called a Ferguson but eventually relented and it was called a Ford, with the familiar blue oval badge fixed centrally above the grille. As a concession, beneath it was another badge

ABOVE: Ford 9N and single furrow plough at a working day.

ABOVE: Ford 9N tractor with Ferguson-Sherman single furrow plough.

ABOVE: A hinged flap covers the fuel filler cap(s) on Ford 9N and 2N tractors.

ABOVE: The bonnet on the Ford 9N and 2N did not hinge, allowing easy access to the radiator cap.

declaring 'Ferguson System'. By December 1, some 6849 tractors had been produced. The first 700 used the aluminium grille and bonnet, while those after used a steel grille and bonnet, with vertical elements replacing the chevron ones. By April 1940, production had increased to 300 tractors per day. ●

RIGHT: There's no doubting who built this tractor, the Ferguson name is relegated. Model number is stamped on the Ford badge.

ABOVE: Blackhawk in the US made a range of equipment for the Ford 9N and 2N.

ABOVE: Ford 9N Mototug was popular on airfields.

ABOVE: In many respects the Ferguson TE20 was very similar to this Ford 9NAN.

ABOVE: Twin fuel tanks denote this to be a Ford 9NAN, TVO-fuelled, tractor.

Chapter 7
Difficult Times

Harry Ferguson had achieved his goal of having a tractor containing his ideas and inventions produced by a major manufacturer, but 1939 was not an ideal year to begin production. Although Henry Ford and Harry had a lot in common and shared a business philosophy of high-volume low-cost production, when they first agreed to co-operate, their age gap may have been a drawback. Ford was 75 and had suffered a stroke, and Harry was 55. Ford had already made some well-meaning but disastrous business decisions and the tie-up with Harry Ferguson would prove to be another – but not immediately.

War in Europe led to supply rationing and in 1942 a pared down Ford tractor was introduced. Labelled 2N (2 for 1942 and N for tractor) it featured steel wheels both front and rear to save rubber, which had to be imported, and dispensed with the electrical system to save the copper used in the dynamo and starter motor. No longer needed, a blanking plate fitted over the hole where the starter was fitted. The battery was also dispensed with, as was the ammeter and its locating hole on the dash panel blanked off.

Starting was now manual by means of a cranked handle, reminiscent of the English Fordson N, locating with the end of the crankshaft and spring loaded to keep it out of engagement when not in use. The knob that operated the choke was relocated to a position on the left-hand side of the front grille for easy adjustment when starting. The ignition spark was now produced by a magneto and fitted by means of an adaptor into the position vacated by the distributor. In place of the dynamo, an idler wheel fitted to a bracket allowed adjustment of the fan belt tension and eliminated the need to specify another size belt. After about 18 months, restrictions eased and the 2N reverted to a similar specification to the 9N.

In 1942, Henry Ford's son and likely successor, Edsel, died, leaving something of a gap. Ford was by now 80 years old and had suffered another stroke. Edsel's son, Henry Ford II, was recalled from the navy in 1943 and took on the role of vice-president at the age of 26. It wasn't long before he realised that Ford Motor Co. wasn't in good shape and some of the problems were caused by the tractor side of the business, which had run up losses of $25,000,000 since production began.

It has been said that Ford hadn't truly costed the production of the Ford 9N and 2N tractors properly and, keeping to his philosophy of producing low-cost products, merely asked which was the cheapest tractor on the market – the Allis Chalmers model B, which was nowhere near the Ford tractor specification – and just undercut it price-wise. Hardly a recipe for success.

Henry Ford II wanted Harry Ferguson to enter into a proper written agreement but without success and after two years of trying he decided to end 'The Handshake Agreement' on November 11, 1946 – but would continue to supply tractors to Ferguson until June 30, 1947. During those months Dearborn Motors was formed, which took tractors from Ford Motor Co. and resold them to Harry Ferguson for a mark-up. With Ford supplying tractors to their own dealers after June 30, 1947, Ferguson only had implements to sell and its dealer network collapsed. Later Ford launched a new tractor model, the 8N (8 for 1948 and N to denote tractor in Ford parlance), a development of the 9N and 2N which looked like a cross between a 9N and a TE20.

The gearbox now had four speeds and the colour had been changed to a light grey for

ABOVE: A Ford 8N tractor that was at the centre of the court case between Ferguson and Ford.

the bonnet and mudguards, with the Ford script embossed into the bonnet sides and mudguards and highlighted in red paint. To further distinguish the tractor, the engine and transmission were painted red.

Although any agreement between Ford and Ferguson no longer existed, the 8N tractor still used the Ferguson system of hydraulic draft control, which was an infringement of Ferguson's patents. Although the ending of the agreement for Ford to produce tractors for Ferguson was said to be mutually agreed, Harry sued the Ford Motor Co. for several millions of dollars, claiming that the firm had deliberately put him out of business.

Henry Ford II tried to settle the case out of court and went as far as travelling to England to meet with Harry to make an offer of $10,000,000 in compensation. This was rejected and the case continued. After two years, whether through boredom or fear of losing the case, Harry sought an out-of-court settlement and was awarded $9,250,000, some $750,000 less than he was originally offered. Although he had won a moral victory, lawyers' fees and other costs used up most of the money. •

Chapter 8
Ferguson Tractor Building in England

Harry Ferguson wanted his new tractor to be built by the Nuffield Organisation, run by William Morris, Lord Nuffield. It is thought that Morris was also keen to build a tractor using Ferguson's hydraulic system, despite having failed to take up manufacture of the Model A. Possibly the success of the 9N and the slow march of mechanisation on British farms had changed his mind.

Yet again no agreement was reached between the two men; it is believed that Morris disliked Ferguson's sales pitch, of which he answered each question by starting again from the beginning of a monologue. Morris is also said to have disliked the model Ferguson used to demonstrate the principle and advantage of ploughs linked to the tractor at three points versus trailed implements. This seems strange since the model was a very effective way of demonstration and would eventually be used by Ferguson company salespeople well into the 1950s.

Contact was made with Sir John Black, managing director of the Standard Motor Co., based in Coventry, who was anxious to get involved with tractor production, initially sending a representative to the US to look at the Ford tractor and see if

ABOVE: A 1950s version of the clockwork demonstration model of the Ferguson System provided to individual salespeople.

Ferguson's patents could be circumvented which, of course, they couldn't.

Black was persuaded to travel to London to meet Harry Ferguson on August 3, 1945, and during the meeting expressed his concern at a possible lack of steel to build the tractors. Harry arranged a demonstration of the tractor (presumably a 2N or 9N) on August 21 and present were Black and Stafford Cripps, Chancellor of the Exchequer, who was impressed enough to promise that sufficient supplies of steel would be forthcoming. During a dinner on September 5 at Claridges, where Ferguson was staying, Black agreed to proceed with the production of the tractor. Such was his standing among board members, combined with Harry's insistence, this was another 'gentlemen's agreement'.

Premises in Banner Lane, Coventry, had been used as a shadow factory to build Bristol aeroplane engines. With the war over, Standard obtained permission from the Government to retain the site's use. The original intention was to transfer car production there from Canley, but Banner Lane proved ideal to build the tractor, as Harry had already discovered, and the move from Canley never took place.

The initial enthusiasm soon soured. Standard was expecting to build a tractor exactly like the Ford 9N, but that was a Ford, designed by Ford and using many Ford parts. Although Harry assured Standard that he had full working drawings for the new tractor and would provide a prototype, in reality this was not the case – the drawings were incomplete and illegible, and the tractor provided only a mock-up. As much of the tractor would be new, it was left to Standard engineers to produce the designs. Although the finished article had more than a passing resemblance to the Ford 9N, the grille and bonnet looking almost identical to that fitted to the first 700 tractors. Where the

ABOVE: Harry Ferguson and Sir John Black of the Standard Motor Co.

9N and 2N bonnets were fixed in position with the radiator cap above the bonnet and an access flap for the fuel tank cap, the bonnet of the new tractor hinged forward to allow access to the battery, fuel tank and radiator. Other features that Harry would have liked to incorporate into the Ford tractor – an overhead valve engine and a four-speed gearbox – were part of the design of the new tractor.

It wasn't only the lack of a design and drawings that annoyed the Standard people. When the Standard-built engine, which had been designed for the new Vanguard car, was announced, Harry kept referring to it as the Ferguson engine. He was also prone to making speeches stating the tractor could eventually be produced for just £100; the reality was more than three times that amount.

After a period of time, Ferguson agreed he had caused delays to the production of the tractor and would pay compensation to Standard, the amount spread out by being paid per tractor produced rather than in a lump sum. In 1947 there was more disagreement when tractor sales slowed due to currency problems and Standard wanted to turn part of the Banner Lane factory over to car production. Harry accused Standard of using high tractor prices to finance car production and gained a £20 reduction in the cost of each tractor. A devaluation of the pound of some 30 per cent boosted tractor sales.

For all the frustrations of the arrangement, the production of tractors for Ferguson proved to be very beneficial to the Standard Motor Co. In 1948 more tractors than cars were produced, and the following year the plant built more than three times the amount of tractors Ford did at its Detroit plant.

Not only was production up – so were profits from the tractor side. In 1951, 60 per cent of the Standard Motor Co.'s profits came from tractor manufacture, and the next year this increased to 70 per cent.

However, the disagreements between Harry Ferguson and Sir John Black resulted in the two companies never becoming fully integrated. The shock news in 1953 of Ferguson selling out to Massey Harris led Black to vow not to produce any more Ferguson tractors when the existing agreement ended. Standard, he said, would build its own tractor and the company went as far as building a prototype, which still exists. In the end, a complete reversal of this decision took place and a new 12-year agreement between Standard and Massey-Harris-Ferguson, as the new company was initially known, was signed. ●

LEFT: Awaiting despatch, Ferguson TE20 tractors at the Banner Lane factory.

Chapter 9
Tractor England

The tractor built at Banner Lane, Coventry, was designated TE20 – Tractor England, 20 horsepower – some 85 per cent of the 24hp it produced at maximum power. To the casual observer it would look like an early Ford 9N.

As the Standard engine designed for the new Vanguard car was not yet ready, the tractors used an imported four-cylinder overhead valve engine built by Continental Motors, of Muskegon, Michigan, designated Z120 and with a capacity of 1966cc running on petrol at a compression ratio of 6:1. The engine was essentially a car engine and required some modification for use in the TE20, as many tractors were built on the monocoque system, including the Ferguson A and Ford 9N, whereby the various castings are load-bearing and no separate frame is necessary.

The Continental had a pressed steel sump which was unable to take any loading, so two tie bars, one each side of the engine, connecting the gearbox bell housing to the front axle support bracket, took the load. The exhaust manifold had its outlet at the rear, requiring a kink in the exhaust pipe so it cleared the gearbox casting. A governor mechanism was added to maintain constant engine speed without driver intervention.

As Ferguson had desired, the tractor featured a four-speed gearbox that incorporated something of a novelty. To start the tractor the gear lever was moved forward and to the right where it engaged with an internal solenoid that sent electricity to the starter motor. This arrangement prevented the accidental starting of the tractor while in gear.

One of Ferguson's early aims was to produce a tractor that weighed no more than

ABOVE: Because it used a pressed steel sump, the Continental engine in the Ferguson TE20 required bracing bars to be fitted.

ABOVE: Having an exhaust manifold with a rear exit required a kinked pipe on the Ferguson TE20.

9. TRACTOR ENGLAND

two draught horses. The Ferguson A weighed 16.5 cwt, whereas the TE20, a larger and more powerful tractor, weighed just over 22 cwt; extensive use of aluminium alloys kept the weight to an acceptable level. Although the detailed design of the TE20 was carried out by Standard staff, most of the basic dimensions were the same as the Ford 9N to save time during development.

The first tractor rolled off the assembly line in 1946 and during that year more than 300 were produced. The next year the figure exceeded 20,000. To cater for working in more confined spaces a narrow version was introduced in 1948 and designated TE B20, with the width of the track reduced to 32 inches. To reduce the tractor to this width, new, narrower axle housings were made, the front axle reduced in width and the radius arms, drag links and drop arms much modified to work with the narrower axle.

In September 1947 the new Standard engine, designed for the Vanguard car and suitably modified for the tractor with a cast aluminium sump and engine governor, was fitted to production tractors, which were designated TE A20. The new engine had a slightly smaller capacity than the Continental engine at 1850 cc but was rated at the same horsepower. Tractors with both the Continental and Standard engines continued to be produced alongside each other until July 1948 when production of tractors with the Continental engine finished. The Standard engine eliminated the need for the tie bars and the kink in the exhaust pipe.

One feature that stood out was the external oil filter, fitted to the left side of the engine. Initially this was fitted vertically downwards but because such a large unit could be prone to catching on obstructions, on later engines it was repositioned to sit at a rear-facing angle to avoid this. Other modifications were introduced with the TE A20. The pads of the individual rear brake pedals were made larger and the centres of the front and rear wheels were now of a scallop design rather

ABOVE: The oil filter on early versions of the TE A20 went vertically downwards, making it vulnerable to damage.

than circular, making track adjustments to the rears much easier. The narrow version of the tractor was now designated TE C20 and in 1952 a vineyard version of the tractor was introduced, designated TE K20, which besides the narrow width was also lower in height due to the fitting of smaller wheels and tyres. An industrial version of the tractor was available and designated TE P20.

No doubt due to the engine now being produced in the same works as the tractor rather than being imported, the price of the Ferguson was reduced from £343 to £325.

In May 1949 after bowing to pressure, a TVO-powered tractor was introduced and produced from July that year. Designated TE D20, the tractor had a larger engine capacity of 2088cc than the TE A20, achieved by widening the diameter of the cylinder bores by 5mm. This was to compensate for the loss of power due to running a lower compression ratio to enable TVO to be used without detonation in the cylinders. The cost was £335.

The fuel tank now had two compartments, the smaller one holding one gallon of petrol used in starting the tractor and the larger one holding seven gallons of TVO. A temperature gauge was fitted to the left-hand side of the dash to indicate to the driver when the engine was warm enough to switch over from running on petrol to TVO, carried out by moving the three-way fuel tap to the appropriate position.

Up until September 1949, commercial petrol, which carried a red dye to distinguish it from ordinary petrol, was untaxed. After that date all petrol was taxed, making running a petrol-powered tractor more expensive and giving farmers more reason to operate tractors powered by a TVO-burning engine.

The narrow tractor with TVO engine was now designated TE E20, the vineyard model TE L20 and the industrial version TE R20. A range of tractors were produced for overseas markets that ran on lamp oil (as the name suggests, an oil produced for burning in lamps with a wick to produce light). Lamp oil is like paraffin but requires an even lower compression ratio than TVO for the engine to run correctly and is widely available in some countries. As with TVO tractors, petrol was

ABOVE: A narrow TE E20 sits alongside two agricultural versions, the TED 20 on the right fitted with optional lighting.

9. TRACTOR ENGLAND

required for starting and getting the engine up to temperature before switching fuels. Tractors fuelled by lamp oil were designated TE H20 for the normal tractor, TE J20 for the narrow version and TE M20 for the vineyard version. Only one industrial lamp oil tractor was built.

In 1951 a diesel-engined version of the Ferguson tractor was introduced, using the Freeman Saunders-designed engine built by Standard and designated TE F20. The capacity of the engine was only fractionally larger than that of the TE20 and the TE A20, at 2092cc, and produced similar horsepower, the benefits being more torque (pulling power) and much better fuel economy. Also, the red diesel used in agricultural tractors was free of duty. Diesel fuel is heavier than TVO and doesn't vaporise, instead relying on a high compression ratio to heat the fuel to burning point. The high compression requires a much stronger engine to withstand the forces generated, which in turn makes a diesel engine much heavier than the equivalent petrol or TVO versions.

To aid starting, particularly in cold weather when lead-acid batteries are less efficient, the TE F20 was modified in several ways. The electrical system was up-rated to 12 volts, with two large six-volt batteries positioned either side of the transmission housing and next to the mudguards and linked in series. The greater electrical power was necessary for the starter motor to spin over the engine. As further aids to starting, decompression levers were positioned at the front of the engine and under the dash panel and when engaged the engine could be turned at a higher speed and would start when the levers were released… hopefully.

First used on aircraft engines, a system built by Ki-Gas of Warwick was a further aid to cold starting. Fuel from a small auxiliary tank was sprayed into the inlet manifold by means of a dashboard-mounted hand pump, the spray heated by a glow plug in the inlet manifold operated from a switch on the dash panel. Perhaps rather surprisingly, given the extra effort required to start a diesel engine, the TE F20 was supplied with a starting handle and it is rumoured that a fitter who worked at a Ferguson dealer was strong enough to be able to start the tractor by this method.

The narrow and vineyard versions of the Ferguson tractor were not produced in diesel form; only the agricultural tractor and the industrial version, designated TE T20. In 1956 an enlarged version of the Standard diesel engine was introduced and designated 23C. As opposed to the 20C original, the 23C engine went into a new tractor, the 35, introduced in 1956, which had improved transmission. In a departure from the all-grey colour of the previous models, the engine and transmission were now painted a metallic bronze. ●

ABOVE: Doing what it was designed for – a TE F20 ploughing.

RIGHT: A Ferguson TE T20 with Scottish Aviation cab has entered into preservation after years of work.

RIGHT: Industrial versions of the Ferguson tractor came with varying levels of equipment.

BELOW LEFT: Industrial versions featured hydraulic brakes and a handbrake.

BELOW RIGHT: Most nuts on the Ferguson tractor were of two sizes, which this spanner would fit (inch scale is to assist in adjusting ploughs).

Chapter 10
Ferguson Tractors Overseas

With the end of the agreement between Harry and Ford for the company to supply tractors to Ferguson, alternative arrangements were needed if he was to continue to supply the US market.

Although Harry had never intended to be a manufacturer of tractors, merely an inventor and sales organisation, he was unable to find an American company to take on the US manufacture of the tractor built by Standard in England, and as a result resorted to setting up manufacturing facilities himself.

A largish piece of land near Detroit was purchased, and the architects were instructed to design a modern-looking building and get it built to a tight timescale. The plot was purchased in January 1948 and in June the building was almost complete, with just a few more weeks needed. On October 11 Harry was able to drive the first completed tractor – a TO 20 (Tractor Overseas), which used the same Continental engine as the TE20, with major components imported from England and smaller items and electrical equipment later sourced in the US.

To distinguish the TO tractor from Ford's latest offering, the 8N (which was painted light grey with the engine and transmission

ABOVE: Ferguson tractors in France were originally built by Hotchkiss.

in red), the TO tractors had a grey bonnet and mudguards, and a dark green engine and transmission. As many of the implements used with the 9N were not supplied by Ford, there was no disruption to this side of the business; many of the US implements for the TO 20 identical to those built for the TE20. The US market for tractors proved to be very competitive and although some 60,000 TO 20 tractors were built up to August 1951, in the same period Ford produced 420,000 8Ns. With an increase in both engine capacity and horsepower, an up-rated version of the TO tractor was introduced in August 1951. With strengthened transmission components it was designated TO 30 to reflect the increased power.

Farms in France tended to be small and family-run, making the Ferguson TE20 an ideal farm tractor. Import licences were difficult to procure, so tractor kits were sent to France to be assembled there and eventually tractors were produced under licence by Hotchkiss at a factory at St Denis, in the suburbs of Paris.

The first of the French-built Fergusons was designated FF 30 and available in petrol and diesel versions, both producing a little more power than the UK versions due to

ABOVE: A sprung seat is a feature of French-built Fergusons.

ABOVE: French-built vineyard Ferguson.

ABOVE: Identification plate on an early French-built tractor.

10. FERGUSON TRACTORS OVERSEAS

ABOVE: Later French-built tractors were distinguished by a grey and red colour scheme.

some tweaking of the engine settings. The ancillary items were French made, and another departure was the fitting of a sprung driver's seat. Narrow and vineyard versions were produced alongside the standard tractor. To distinguish French-built tractors, the engine and transmissions were painted bright red with the bonnet and mudguards in the familiar grey. With tractor production in France amounting to 20,000 units per annum, the St Denis factory was at its production limit and a new factory was built at Beauvais, some 40 miles away. Tractor kits were also sent to India and Australia for local assembly. ●

Chapter 11
Ferguson Conversions

Although the Ferguson TE tractor and its range of implements were universal in their applications, there were still areas where they did not exactly meet users' requirements and a number of outside companies either modified existing components or produced new parts.

Harry Ferguson was always a believer in petrol-powered engines (gasoline or gas in US terminology) which was acceptable where petrol was not expensive, but in the UK petrol was somewhere around twice the price per gallon of paraffin or tractor vaporising oil (TVO), which many contemporary tractors used, and both the TE and later TEA were only available with petrol power. Several companies spotted this gap in the market, notably Vapormatic, of Budleigh Salterton in Devon, and Norfolk-based Loddon Engineering, and produced kits to enable Ferguson tractors to be used with the cheaper fuel.

TVO has different characteristics to petrol. It requires a higher temperature than petrol to vaporise and be burned in the engine cylinders at a lower compression ratio, and the conversions controlled these. Petrol was still required for starting and running the engine up to operating temperature when the changeover

ABOVE: A number of companies offered TVO conversions for the Ferguson TE A20. This one is by Loddon Engineering.

to running on TVO could be made. An extra small tank was required to hold about one gallon of petrol and the former larger petrol tank given over to holding the TVO. To assist in the TVO warming process, a new thermostat was fitted, and the main part of the conversions was a new inlet and exhaust manifold that utilised hot exhaust gasses to vaporise the TVO. To lower the compression ratio, spacers and extra gaskets were added to the joint between cylinder head and engine block. A tap with three positions (petrol, off, TVO) was required to connect the tanks to the carburettor, which also needed modification, mainly the addition of a drain screw in the bottom of the float chamber to release the TVO used in the previous running and allow the chamber to be filled with petrol ready for starting.

Ferguson introduced its own TVO conversion for the TEA, along similar lines to the proprietary ones. The toolbox was relocated on to the left-hand mudguard and in its place alongside the battery fitted a small rubber-mounted petrol tank. The manifolds were encased in a distinctive bare metal aluminium shroud, designed to keep in more heat, which would be carried over to the production TVO tractor.

ABOVE: Ferguson factory TVO conversions and TE D20 tractors are distinguished by the aluminium manifold cover.

The official conversion was a lot more complex than the non-factory ones, in that it required complete dismantling of the engine and the cylinder block bored to take larger liners and pistons of 85mm diameter as opposed to the original 80mm diameter, bringing the engine to very similar specification as the later TED. Modifications to the distributor were also required – a 20-page instruction book detailing all the stages of the conversion.

Although the Ferguson three-point linkage transferred weight to the rear wheels of the tractor giving increased grip, one company thought the tractor would benefit from even more grip in the form of a four-wheel drive system. Selene of Italy had carved out a niche converting virtually unsellable Fordson E27N tractors into four-wheel drive tractors.

By modifying ex-military GMC axles and fitting them in place of the standard front axle, drive was taken by propeller shaft from a transfer box fitted between the transmission cases. A similar

ABOVE: For TVO conversions, a separate petrol tank and three-way fuel tap were required.

ABOVE: Selene 4WD conversion used an ex-military Jeep front axle.

ABOVE: The Selene 4WD conversion was not hugely popular.

system was developed for the Ferguson, using Jeep front axles fitted with extra brackets, again the drive was taken from a transfer box fitted between the gearbox and axle casings. Very few of these conversions were produced and even less survive, as they negated two of the Ferguson's characteristics – a tight turning circle and an adjustable front axle.

Perkins, of Peterborough, started producing diesel engines in the mid-1930s and had gained a reputation for reliable, economical engines fitted to a variety of mainly commercial road vehicles, where the manufacturers did not produce their own diesel engine. Ferguson was a late convert to the advantages of diesel power and Perkins offered a diesel conversion kit for the TE20 tractor long before Ferguson offered a factory-produced diesel option.

The Perkins conversion consisted of the three-cylinder P3 engine which gave far greater horsepower than the petrol engine it replaced and could also be run on the very much cheaper 'red' diesel, which, because tractors were used off-road, carried a much lower fuel duty than the diesel used by road vehicles. With the P3 being a tall unit, the tractor bonnet had to be raised to clear the top of the engine and the radiator was also raised, resulting in a somewhat ungainly look to the tractor.

In Scotland, Gavin Reekie graduated from general agricultural repairs to become a Ferguson dealer in Arbroath, an area known for berry growing, particularly raspberries. It would be the raspberry growers who went to Gavin Reekie to produce a tractor suitable for use between the rows of canes and obviously, being a Ferguson dealer, this was the tractor he chose to modify.

Narrowing the Ferguson to meet the growers' requirements took some drastic action. The rear axle housings were cut through and several inches of material removed. The two halves were then welded back together with reinforcing pieces attached. Material was cut from the front axle and the drop arms, with drag links and radius arms bent to bring them into line with the narrower front axle. The rear wheels were reversed, and the mudguards rotated forwards to provide protection for the driver from the now much nearer rear tyres, the modifications bringing the width of the tractor down to 44 inches but making it a little unstable. The cost of the conversion alone was £250, almost as much as the original tractor price of £325. Harry Ferguson was not impressed with this somewhat brutal conversion and tried to buy out Reekie, who refused even though he was a Ferguson dealer and could have had the franchise taken from him… another case of two independent men meeting head-on.

Harry would go on to build factory versions of narrow tractors, which were more elegant to look at and also more stable. It is believed about 200 Reekie conversions of the Ferguson tractor were produced in all, with Gavin Reekie fitting his own badge to the bonnet above the Ferguson badge and many sporting a distinctive blue band around the bonnet. A large number of Reekie-converted tractors have survived, and many are now in tractor collections. •

ABOVE: The higher bonnet line of a Perkins P3 conversion allows for advertising space.

ABOVE: Many of the Reekie narrow conversions were distinguished by a blue painted band on the bonnet, as well as the Reekie badge.

ABOVE: An unusual conversion. This is a golf course tractor.

Chapter 12
To Diesel with Reluctance

As with tractor vaporising oil-powered tractors, Harry Ferguson was a late convert to tractors powered by a fuel other than petrol. He believed this was superior, which was probably acceptable in the US, but in the UK, farmers wanted pulling power and economy – and diesel would give them both.

In 1950 testing began using three Ferguson tractors, each fitted with a different make of diesel engine: a three-cylinder example from Perkins, the P3; a four-cylinder unit manufactured by Meadows of Wolverhampton specifically for the evaluation; and a unit manufactured by the Standard Motor Company, the four-cylinder engine designed by Arthur Freeman-Saunders and intended to be used in the Standard Vanguard, saloon, van and pick-up. The tractors were put through identical tests, which besides ploughing also involved towing heavy loads on the road. During the winter, they were left outside overnight to test their cold-starting capabilities. It is most likely that the test of the three engines was not to determine which diesel Ferguson should choose to be manufactured for the tractor, but to provide a meaningful comparison under the same conditions.

Although the Perkins produced more power, the tall unit not only looked ungainly installed in the tractor, it *was* ungainly. Tester Nigel Liney overturned the test tractor on one occasion, thankfully without injury. Notwithstanding the aesthetics, it was also likely the Perkins engine would be more expensive to purchase.

There's a possibility the Meadows engine, from an established diesel engine manufacturer, was made just for the test (although it is believed

ABOVE: Exhaust outlet on the left side is a feature of the Perkins P3 engine.

ABOVE: Fuel pump of the Meadows Ferguson sits on the opposite side of the engine to the Standard engine.

ABOVE LEFT: Meadows diesel engine sits slightly higher in the tractor.

ABOVE RIGHT: Unlike the Standard engine, the starter motor is on the left on the Meadows engine.

ABOVE: Tractors with Perkins P3 engine conversions needed the bonnet to sit higher.

several were produced). Being four-cylinder, it was a good comparison against the also-four-cylinder Standard unit, but it was likely that this also would be an expensive engine to produce, as Meadows was more of a high-quality low-volume manufacturer.

Unsurprisingly, the engine used in the production Ferguson tractor (the TEF) was the one built by Standard, a firm which no doubt had the capacity to build in sufficient quantities to satisfy Ferguson's requirements at an attractive price. For whatever reason,

12. TO DIESEL WITH RELUCTANCE

ABOVE: Amazingly, both Nigel Liney and the tractor were unscathed after this accident. *Nigel Liney.*

the starter motor of the diesel fitted on the right side of the engine rather than the left side of the TVO unit, which seems a little strange. The Standard diesel produced slightly more horsepower than the TVO-fuelled engine but not as much as the P3, its advantage being the greater torque and better fuel economy. No figures exist for the Meadows diesel.

Many, if not most, experimental vehicles end up being destroyed by companies after they are finished with, very often to prevent them getting into the wrong hands. Of these three experimental diesels, the fate of two is unknown but the Meadows found its way into private ownership and was used by a farmer for several years before the engine was destroyed through lack of oil.

It lay derelict on a Warwickshire farm before being rescued in 2003 and restored by engineer David White. He salvaged and repaired the broken engine parts and repainted the tractor in a distinctive light green colour, traces of the original colour having been found on unexposed parts of the tractor when it was dismantled. The tractor is now in the collection of Michael Thorne. •

ABOVE: The Meadows-engined experimental tractor stands out with its distinctive pale green paintwork.

Chapter 13
The Ferguson System

As we've seen, Harry Ferguson's foremost concern was to improve farming efficiency and the linked implements with depth control was the way he sought to achieve this. The production of a lightweight tractor came about because nothing suitable was available. Buying just the tractor fitted with the hydraulic system only was of no benefit to farmers; they needed to purchase at least a plough to go with the tractor and ideally other implements as well. This extra expenditure held back sales in the early years, and it was into the late 1940s and early 1950s before the Ferguson System, as it was termed, was universally appreciated.

ABOVE: A 1950s brochure lists the many accessories.

13. THE FERGUSON SYSTEM

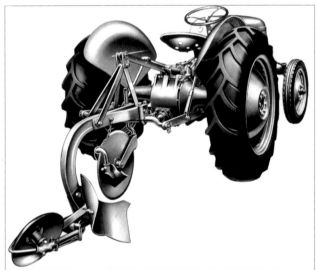

ABOVE: The Ferguson two-furrow plough was the first implement to be developed.

LEFT: For heavier soils, a single-furrow plough was used.

trailed implements to be connected to the tractor. The arrangement used to connect implements to the tractor was first used on the Model A and was so well thought-out that it continued with the 9N and on to the TE20 and later derivatives.

With the duplex plough fitted to the Fordson F there was only one implement, so the need for a quick changeover was irrelevant, and some of the attraction of the Ferguson A and its implements was that they could be changed easily and quickly. The simple layout consisted of short pins on either side of the implement that fitted into articulating part spheres located on the end of each lower link. The pins on the implement could be easily fitted into the spheres

The Ferguson Model A was initially offered with just four implements: a two-furrow plough, a three-row ridger or bouter, a general cultivator or scuffle, and a row-crop cultivator. Later a single-furrow plough was added, together with a drawbar to allow

ABOVE: The two-furrow plough is an essential part of the Ferguson System.

ABOVE: The basic design of the two-furrow plough didn't change. Here is an example for the Ford 9N sits alongside one for the TE20.

13. THE FERGUSON SYSTEM

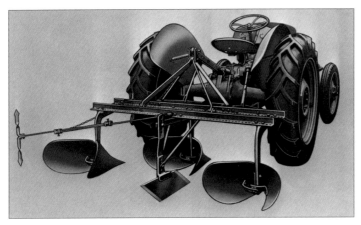

ABOVE: Used mainly for potato crops, the ridger or bouter.

and to keep them in place, link pins with an over centre spring clip were positioned into holes drilled into the implement pins. The spring clip, when positioned correctly, prevented the implement from coming apart from the lower arms.

The implement top link had similar articulating part spheres at each end and joined implement to tractor by pins placed through the top of the implement and a lug on the end of the depth control arm. Again, these pins were kept in place by link pins. So successful was this system that it is the standard by which all linked implements are affixed today, and the diameter of the implement pins are now standardised into four sizes, termed categories, which run from 0 to 3, with 0 being the category for implements fitted to garden tractors and 3 being the one for those fitted to 100hp-plus tractors.

When the Ford 9N was introduced, much was made of the fact that the implements were wheel-less, and the term 'flexible farming' was introduced. A logo box on literature had 'Ford Tractor' as its heading, where the 'T' was similar to the 'F' of Ford but without the crossbar. More importantly, beneath it was the now-familiar image of a tractor and raised plough in profile, the image running left to right rather than the later right to left image and the words 'Ferguson System' super-imposed.

A sales brochure of 1940 shows the following implements: a two-furrow plough, cultivator, row crop cultivator and an implement called a middle buster, which carried one or two bodies and was very similar in design to the ridger but with flatter bodies.

By late 1945 the range of implements available had increased considerably and included the following: a weeder, a tiller with nine tines, a narrow frame tiller with seven tines, a rear mounted mower (termed heavy duty), disc plough, blade terracer, spring tooth harrow, trailed disc harrow, a two-row drill planter, disc terracer, harrow, soil scoop, cordwood saw, a hay sweep, lister planter (planting into a deep furrow) and a crop cultivator. Onto most of these implements was affixed a small plate with 'Ferguson' as the heading and 'Ferguson-Sherman Mfg Corp., Dearborn, Michigan' underneath. Also listed would be the part and serial numbers and the US patents that applied to that implement. This method of identifying implements was carried over to the British-supplied ones.

Both oversized tyres and dual rear wheels could be specified for operation in certain soil conditions. For wheel removal, a high clearance jack could be used on the front or rear axle of the tractor. An adaptor fitted into a spark plug socket used engine compression to inflate tyres to the correct pressure, with the complete kit including a hose, connector and pressure gauge. Two engine fans were listed: a four-bladed winter fan was designed to direct warm air back to the driver in cold weather; while a six-bladed fan was available to use in hot dusty conditions. Instead of drawing air through the radiator from the grill, it pushed air in the opposite direction to prevent the build-up of dust and small particles in the radiator core.

ABOVE: It was recommended that the rear wheels were set wider when using the earth leveller and blade terracer.

RIGHT: Official Ferguson implements were identified with a brass plate.

13. THE FERGUSON SYSTEM

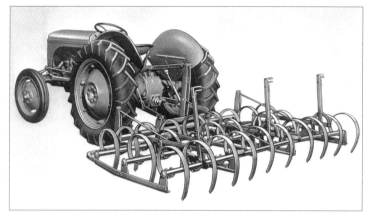

LEFT: Spring tooth harrow was used for seed bed preparation.

LEFT: The three-gang spring tooth harrow was useful for seed bed preparation.

BELOW: The earth scoop had many uses around the farm, the contents tipped out using a lever accessible from the driver's seat.

A full lighting kit consisting of two headlamps, a tail-light, and associated wiring and brackets was available for after-dark work. One of the headlights could also be fitted to the rear mudguard to be used as a work light. A front bumper could be fitted with brackets to the front axle and was described as offering a wide range of protection. Also available was a lower guard which fitted to the bumper. Its role was to hold down corn stalks and heavy weeds as the tractor passed over.

The storm cover provided protection from the elements and ran from the front of the tractor back to cover the driver's seat. It was made of heavy material and carried the Ford and Ferguson System logos on its sides.

Further implements of a more industrial nature could be fitted to the 9N, many of them manufactured by Arps Corporation of Wisconsin and sold under the Blackhawk name. A 5ft angledozer could be used in conjunction with a mid-mounted grader blade. Fitted to the rear of the tractor on the three-point linkage was the pick-up scraper, a device where the front edge engaged with the ground to pick up earth and when full, the earth could be transported and dumped by opening up the rear of the scoop. For snow-clearing duties a straight front blade was offered, and for heavier conditions was a vee-type front snow plough. All of these attachments used the same sub-frame for fitting them to the tractor.

Looking extremely large and appearing inappropriate for a lightweight tractor was the heavy-duty winch, attached to the front of the tractor and driven from the PTO shaft.

Definitely designed for industrial applications was the Moto Tug, where extensive sheet metal covered the front of the tractor, swept back around the sides and over the front wheels to meet with the wide rear mudguards. Road going-type wheels and tyres were fitted: single rears to the 2,500lb drawbar pull version and twin rears to the 4,000lb drawbar pull version. The Moto Tug was used extensively on aircraft-towing duties and about 275 were built.

With the building of the TE20, the range of implements increased to cope with the different farming tasks. As farmers got more used to the benefits of implements designed for the Ferguson tractor, which was well-suited to British farms, its manoeuvrability proved an asset when working in relatively small fields.

To the casual observer the implements made for the Ferguson A would most likely look the same as those made for the 9N and TE20, because the design differences were only subtle. The original concept and design were so correct that no further development was needed.

With ploughing a major farming operation, the range of ploughs increased. Two furrow ploughs came in five versions, with soil condition deciding choice. For heavier soils there was a single-furrow version and for lighter soils a three-furrow version, plus a kit to convert a two-furrow plough into a three-furrow plough. A popular accessory for two-furrow

ABOVE: A tractor cover enclosed both the engine and seat, however very few survive.

ploughs was the adjustment handle, which could be used from the driving seat. For certain conditions a single-furrow reversible plough was offered. This had right-hand and left-hand bodies, so all the furrows faced the same way. Upon lifting, the plough swung over to allow the other mouldboard to be used. For hard and dry soil, mainly overseas, a disc plough, with either two or three discs, was offered.

There was a great range of implements available for working the soil after ploughing. Disc harrows were used to break up heavier soil and Ferguson offered several versions, both mounted and, perhaps surprisingly, trailed. The offset disc harrow was mounted on the tractor and the ability to move the discs to one side was an advantage in orchards and vineyards where cultivation under trees and vines was required. The tandem disc harrow was available in 6ft and 7ft widths and also tractor mounted, which proved useful in small fields where it could be reversed into corners. The four sets of discs were set two behind two, and the angle of the discs adjustable by two frame-mounted levers. In modified form it was used in paddy fields, where cage wheels were fitted to the tractor and measures taken to prevent water entering the working parts.

The heavy-duty reversible disc harrow was of a more specialised nature. The two sets of four scalloped discs could be set to either drive soil into the centre, which was useful in seed bed preparation, or drive material such as mulch outwards. A sort-of leftover from the 9N era was the semi-trailed tandem disc harrow, available in both 5ft and 6ft widths. This implement was not of the typical trailed type nor mounted but required its own linkage, which not only towed but also angled the discs, attached to the tractor. An attached skid plate operated the hydraulic sensor to give weight transfer under certain conditions.

Many of the ground engaging implements were constructed along similar lines, again to a design first seen with those built for the Ferguson A. The main structure of the two cultivators and the ridger made for the A were two pieces of steel angle drilled with numerous holes, the triangular frame to facilitate depth control affixed above the two angle pieces.

ABOVE: A kit allowed a two-furrow plough to be converted into a three-furrow version.

ABOVE: The low volume version was one of two sprayers available.

ABOVE: Disc ploughs were mainly used overseas where the ground was hard.

RIGHT: Ideal for use in orchards, the offset disc harrow.

BELOW: The mounted tandem disc harrow was ideal for use in small fields.

13. THE FERGUSON SYSTEM

ABOVE: This disc harrow is for use in paddy fields, where the cage wheels aid traction.

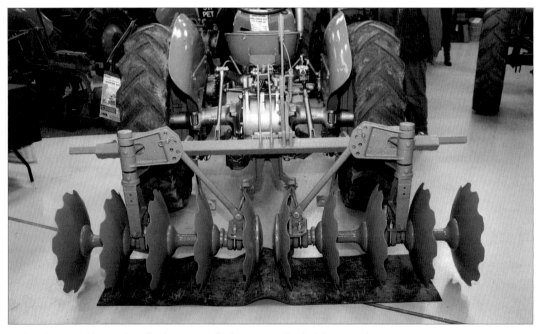

ABOVE: A restored heavy-duty disc harrow on display at an enthusiasts' tractor show.

ABOVE: Originally developed for the 9N tractor, the semi-trailed disc harrow.

The purpose of the holes was to allow tines and ridging bodies and so on to be bolted on at varying distances compatible with the width of the crops being cultivated.

The tiller was a similar device to the general cultivator and introduced alongside the Model A, in that the tines were tensioned with coil springs. This was to prevent damage when an obstacle was struck, the tines being able to move back against the spring tension, which then pulled them back into place when the obstacle had passed. A number of different feet were available for fitting to the tines, depending on the work being carried out. With a similar name but slightly different construction was the spring tine cultivator, where the tines were made of spring steel to allow some flexibility but not as much as the tines on the tiller. This implement was designed for lighter, mainly row-crop cultivation.

Another implement for soil preparation was the Cult-Harrow, built by Horstman of Bath. Driven from the tractor Power Take-Off (PTO), two rows of tapered tines, each containing 12 tines, reciprocated in opposite directions to work the soil. While raised and lowered by the

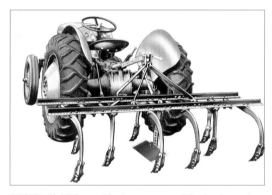

ABOVE: Rigid tine cultivator was one of the implements that used depth control.

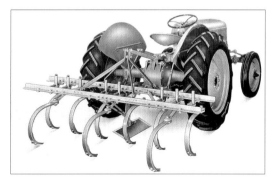

ABOVE: Spring tine cultivator was used mostly for row crop work.

13. THE FERGUSON SYSTEM

three-point linkage, a chain attached to the top link took the weight of the implement. This is another rare Ferguson implement prized by collectors and commands a high price when available.

Built for weeding between rows of crops were two steerage hoes that both used the multi-holed frame. The independent gang version was the slightly more complex version, in that the individual units could move independently of each other. On the more basic version the units were fixed. Both types were 'steered' by an operator sat on a tractor-style seat well to the rear of the hoe, using a horizontal curved bar coupled by link to the right-hand lower link to keep the units between the rows of crops. The hoes on both types were kept in a central position by means of a large tension spring between implement and tractor, the operator being able to move them up to five inches either side of central. When these hoes were first introduced, many tractors were still equipped with the original down swept exhaust, which put the hoe operator directly in line with the exhaust fumes. Swapping the original exhaust system to the Ferguson accessory of a vertical exhaust running up the side of the bonnet cured the problem.

Of an entirely different construction but also used in close crop work was the row crop thinner, whereby rotating arms driven by two land wheels struck out young plants at intervals to allow the remaining plants to fully develop. This implement did not make use of the draft control but saved an awful lot of hand work with a hoe.

Weeds are a constant problem for farmers; they take up valuable nutrients needed by the crop and ploughing the ground and thus burying them is an over-winter solution. Once the crop is growing, other solutions are required and the weeder was one of these. At 13ft wide, it comprised of 71 thin spring steel tines attached to a frame whose ends could be folded for transport. The slender tines vibrated when in contact with the ground and barely penetrated the soil. A top link balance spring offset some of the weight of the implement and in row crop work some of the tines could be removed. Chemical weed killers later took on the role of this implement but the weeder continued to be used on organic farms.

Potatoes are a major food source and the Ferguson system catered extensively for both its planting and harvesting. The ground preparation for planting would be the same whichever planter was used. After the soil had been cultivated and a good tilth obtained, ridges would be formed using the three-row ridger, where the central body was located forward of the outer two. Two planters were initially offered, and these were sold as kits to be fitted to the ridger. The planter for non-chitted potatoes consisted of a large centrally situated hopper. The internal baffles allowing only a few potatoes at a time to roll onto a tray in front of each of two operators, who sat on tractor-style seats sideways to the implement on the left and right. The seed potatoes would be manually placed into tubular chutes which dropped them into the bottom of the ridges. To ensure even planting, a notched ground wheel operated a bell that signalled to the operator when

ABOVE: The steerage hoe required an operator to keep it within the crop lines.

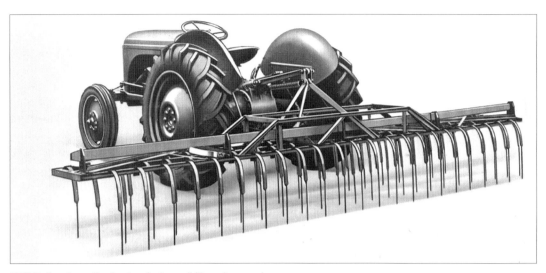
ABOVE: An alternative to chemical weedkillers, the weeder.

a potato should be dropped down the chute. Different planting spacing could be achieved by adjusting the bell mechanism, so it engaged with a different set of notches on the wheel.

Chitted potatoes are those that have already begun to sprout and require careful handling to avoid damaging the chits, and the trays in which they were chitted would be placed on framework on the planter, several being loaded at one time.

Again, planting was manual: potatoes were placed into a chute, their spacing controlled by the bell, but in this instance the two operators sat on seats at the rear of the implement to allow access to the trays. With both types of planter, two rows of potatoes were planted simultaneously while the tractor was driven along the tops of the ridges, and the ridger filled in the planted trench as it went along.

ABOVE: The potato planter for chitted potatoes, which were placed by hand into a chute at the prompting of a bell.

13. THE FERGUSON SYSTEM

ABOVE: Potatoes from the hopper were fed by the operators into a chute.

system was said to be more economical than broadcasting fertiliser after the potatoes had been planted. Later an automatic potato planter was introduced but was only suitable for non-chitted potatoes and was too heavy to be used on the TE20 and its later derivatives.

To plant seeds such as wheat, a universal seed drill was offered, which would plant 13 rows at a pass, the seeds carried in a hopper above, with the mechanisms driven from the wheels of the seed drill. As no great force was involved the tyres didn't require lugs for grip and were merely ribbed. Coupling to the tractor was via a secondary linkage attached to the three-point linkage of the tractor, the raising or lowering of which brought the seeding units in or out of the soil. Accessories were available to cope with different seed sizes and soil types.

To aid growth, a fertiliser attachment could be used with either planter. Driven from an attachment to the left rear wheel of the tractor, this attachment sat above the rear axle with the granular fertiliser in a hopper. A range of various-sized drive sprockets allowed for different amounts of granular fertiliser to be delivered into the trench ahead of the potatoes, and this

ABOVE: A potato planter fertiliser attachment displayed on a restored tractor.

ABOVE: On both of the potato planters, seed potato was fed manually down a chute into the furrow.

To apply fertiliser at the same time as planting, a fertiliser attachment could be fitted to the rear of the seed hopper and also driven via the seed drill wheels.

Fertiliser distribution after seeding when plants were established could be carried out by a spinner broadcaster, which fitted onto the three-point linkage. A hopper held the granular fertiliser that fed onto a rotating plate driven from the tractor PTO, centrifugal force spreading the fertiliser up to a width of 16ft. The rate of flow onto the disc was controlled by lever from the tractor seat.

The two-row planter drill was designed to be used mainly in the US for sowing cotton or maize. The mechanism was driven by chain from land wheels, which also pressed down the soil after the seeds had been planted. Small furrows were formed ahead of the planting process, with fertiliser from separate hoppers being placed into these before the seed was introduced.

For harvesting potatoes, or potato picking as it was usually called, an implement called a potato spinner was used due to the manual nature of the task. Similar in operation to the many trailed versions produced over the years by a number of companies, the Ferguson spinner was, of course, mounted on the tractor and could be raised and lowered hydraulically rather than relying on a wheel-driven mechanical device used on the trailed spinners. As with many of the Ferguson implements, this enabled greater manoeuvrability. The Ferguson spinner took its drive from the tractor PTO rather than a land wheel, which could slip in muddy conditions.

A blade ran along the bottom of the ridge and underneath the crop of potatoes and the PTO powered two reels, spinning in opposite directions, one to loosen the potatoes from the

ABOVE: The broadcaster took its drive from the tractor PTO.

13. THE FERGUSON SYSTEM

soil and the other to cast the leaves and stalks to one side. A canvas curtain prevented the potatoes from being flung onto the adjacent unharvested row. Two additional stays running from the ends of the lower link arms and on to pins bolted to the axle underneath the mudguards prevented any sideways movement, keeping the spinner in line with the ridge. These additional stays could also be used with other row crop implements to keep them in line.

Although somewhat superior in construction, the Ferguson potato spinner still relied on manual labour to pick the potatoes from the ground and place them into conveniently placed sacks. This was once a seasonal job that provided welcome work and income in most villages. Later an automatic harvester was introduced, where, instead of the backbreaking picking, the potatoes were delivered to a rotary table which operators stood around and placed potatoes by hand into sacks.

Also mounted on the multi-hole frame were the beet lifters, devices used for getting sugar beet out of the ground, the two-row version using the full width frame and the single row version using a narrower frame. This was again a case where the lifted beet would need manual intervention to get them from the ground and into a trailer. Another operation would be topping the beet, where the leaves and stalks would be removed prior to lifting, the implement avoiding cutting them manually. Only a single-row version was offered, where a land wheel drove by chain a multi-element feeler wheel, which pulled the stalks against a horizontal cutting knife. The stalks and leaves were discarded or gathered by hand for animal feed. It's probable that not many beet toppers were sold, as they are now very collectable among enthusiasts and command high prices when they become available.

Another Ferguson implement designed to take the drudgery out of a manual task was the kale cutrake, essentially a buckrake with a finger bar mower cutter blade set along its edge, the reciprocating blade driven from the tractor

ABOVE: Making workers' lives a little easier, the later potato harvester delivered the crop to a rotary table for sorting.

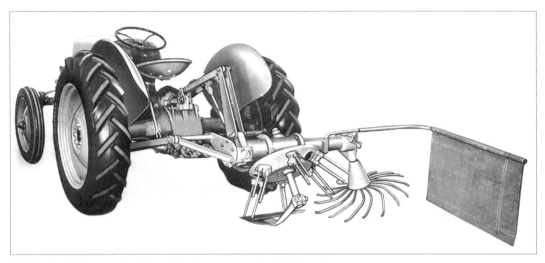

ABOVE: After being uprooted the potato tubers were collected by hand.

PTO via a gearbox. With the blade in operation the tractor was reversed into the kale crop in the lowered position and the cut stalks fell onto the buckrake, which had extended sides to prevent the kale falling off. When enough had been cut, the buckrake was raised for the load to be transported back to the cattle sheds. Surprisingly, given the unpleasant nature of kale-cutting by hand – usually in cold and wet conditions – the kale cutrake wasn't widely popular and so is another sought-after implement that fetches a high price when one comes up for sale.

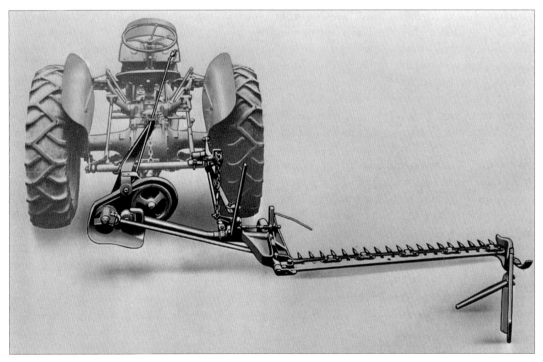

ABOVE: Mounted at the rear of the tractor the finger bar mower required the driver to look backwards to check on progress.

13. THE FERGUSON SYSTEM

ABOVE: Surprisingly, given the unpleasant nature of kale cutting, the kale cutrake was not a popular implement and is now sought-after by collectors.

Other ground-engaging implements of the Ferguson system didn't use the multi-holed angle frames, mainly because as they were not used for row crop work, no adjustment was required. The adjustable spike tooth harrow was attached to the tractor via the three-point linkage but did not use depth control. A rack on the top link connected a tension spring between the link and the harrow, its purpose to prevent excessive digging in the front row of spikes, which were in three sections, the two outer ones folding for transport. The angle of the spikes and, subsequently, spike penetration was controlled by adjusting levers on each section. Although used for similar work, the spike tooth harrow heavy duty was constructed differently. Four sets of harrows were suspended by chains from a tubular framework, with the two outer sets folding for transport. There was no provision for adjustment; a ratchet on the top link controlled the depth of the harrow's soil penetration.

Another implement for soil preparation was the rotary hoe, which wasn't powered but relied on ground contact to rotate four sets of eight spoked wheels that broke up the soil.

Many farms in the 1940s and 1950s were still what is called 'mixed', in that they kept animals and also grew crops, not only for animal feed but as income. Where cattle for either meat or milk production was kept, a lot of manure was produced. Once rotted down, it would be used on the fields as a fertiliser. Ferguson produced a range of implements to handle and spread the manure or muck, as it was colloquially termed. Two manure spreaders were offered, but not at the same time. The initial model was steel-bodied, with the conveyor and spreading reels driven from the land wheels which, as they were driving as opposed to being driven, had the tyres fitted with the tread pattern reversed. The body of the spreader widened towards the rear to allow the manure to move along without jamming. Acid in the manure caused

ABOVE: The outer parts of the spike tooth harrow fold for transportation purposes.

ABOVE: Very few of the steel-bodied manure spreaders have survived, due to corrosion.

13. THE FERGUSON SYSTEM

corrosion to the steel body and a yearly coating of a special paint was advised, but nevertheless many succumbed to rotting and only a few survive. After the amalgamation, a wooden-bodied Massey Harris spreader was modified so it could use the Ferguson hitch. The wood was more acid-resistant and could be easily replaced if needed.

Harry Ferguson's aim of mechanised farming applied equally to menial tasks, so rather than dig out and load manure by hand, two hydraulic loaders were introduced, and Ferguson is recognised as being the first to come up with such devices. The first was simply termed 'manure loader' and consisted of a large fork with eight tines fitted to tubular steel arms that mounted on brackets fixed to the front axle. Two hydraulic rams, each running alongside the engine and transmission, were fed from a pipe underneath the rear axle and these provided the power to raise and lower the loader fork which was mechanically tripped by wire for emptying.

Later a high-lift version, termed the 'banana loader' because of the curved nature of the lift arms, was added. This loader needed a substantial frame fitted above the rear axle to take the ends of the lift arms, while another frame fitted under the tractor midway took the ends of the hydraulic rams. For this loader to work it was necessary for the rear hydraulic arms to be locked into position by the fitting of the automatic hitch used on the trailers. The recommended concrete counterweight sat on the hitch bar, which was initially used to lift it into working position. An interesting feature of the high-lift loader was the hydraulic ram that tipped the fork or bucket when that option was used.

The 30 cwt tipping trailer was a useful item to transport manure out to the fields. Although this was steel-bodied and succumbed readily

ABOVE: Manure loader attached to the front axle via special brackets.

to corrosion, many that have survived into collectors' hands have been extensively rebuilt, and such is the attraction of this trailer that replicas have been made available.

A very useful and adaptable item was the three-tonne trailer, available in both tipping and non-tipping forms. The very early examples had quite an involved system for connecting to the tractor: a wide beam transferring the load to the top of the tractor axle and bars fixed underneath the tractor actually pulling the trailer. Two tubular jacks at the front of the trailer were lowered to the ground when the trailer needed to be parked unused. This system was soon superseded by a beam fixed underneath the trailer and terminated by a ring to allow coupling to the tractor. An automatic hitch connected the hydraulic arms and a pivoting hook underneath the tractor. To connect the trailer, the tractor was reversed up to the trailer, and the hook lowered and engaged with the trailer ring. Coupling was complete when the hook was raised, and the automatic hitch locked in position.

ABOVE: High-lift loader was nicknamed banana loader after the shape of the arms.

ABOVE: A useful combination of a manure fork and 30cwt tipping trailer.

13. THE FERGUSON SYSTEM

LEFT: Two versions of the 30 cwt trailer were produced, tipping and non-tipping.

The trailer wheels were at the rear end, giving maximum weight transfer to the rear wheels of the tractor. The tipping trailer featured a hydraulic ram situated under the middle of the trailer and feed to it was taken via a hydraulic pipe fitted to the transmission case. Initially the front board and sides were detachable and held in place by strakes slotted into the trailer sides. On later versions the sides were hinged, as was the tailgate on all versions. To increase the adaptability of the trailers, extended sides for carrying grain and a different type for carting silage were available.

Where loose hay needed to be carried, hay lades were available. These were racks that fitted to each end of the trailer at an outward angle to hold the hay in place. Mainly for local

ABOVE: Capable of carrying three tonnes, this MK1 trailer has been converted to MK2 specification.

RIGHT: Connecting linkage for MK1 trailers was complicated and soon superseded by the much simpler MK2 version.

BELOW: Jacks came in different versions and used the tractor hydraulic arms to provide lift.

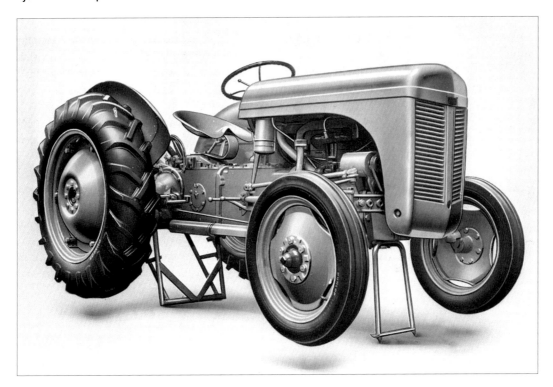

authority use, a version was available with a leaf-sprung rear axle, and mudguards could be specified. For working on soft ground, a kit to convert to dual rear wheels was produced.

The transport box, made of steel, was a popular item for carrying smaller items. It fitted onto the lower link arms and two catches which went over the tops of the arms held it in place to prevent tipping. The sides were triangular and a small hinged flap at the rear was secured in place by two sprung-loaded catches. This item was often used for transporting milk churns and could hold six at a time. With the box on the ground and the rear flap open, the churns could be rolled on edge into the box without the need for lifting.

Many farms were situated down long drives, away from the road, and platforms were erected at the roadside for the churns to be placed on, ready for collection by lorry, with the height of the platform usually the same as that of the bed of the lorry so they could be rolled across without lifting. Empty churns would be left on the stand ready for the next day's milk. The transport box on the Ferguson was ideally suited to the task. An accessory was available to turn the transport box into a wheelbarrow; with the box off the tractor, a tubular frame forming the handles could be fixed to the front edge and a pair of barrow-type wheels fitted underneath, making a useful item for lighter loads. With a larger capacity than the transport box was the transporter, which fitted onto the three-point linkage and was available in tipping and non-tipping versions.

Hay is part of the annual cycle on farms keeping animals, being used as a winter feed when grass is not growing, and the Ferguson System catered well for the processes. To cut the grass, a mower mounted on the hydraulic arms and driven from the PTO was offered. The rotary motion converted to a reciprocating action that drove a multi-sectioned blade along rows of steel 'fingers' which separated the grass stalks prior to cutting. Two widths were offered, 5ft and 6ft. A disadvantage of the mower being mounted at the rear of the tractor was that the operator had to turn around to check that all was well, though mid-mounted versions were offered later.

ABOVE: A handle and wheels converted the transport box into a wheelbarrow.

ABOVE: Ideal for taking milk churns up to the roadside collection point was the transport box.

To avoid wounding or killing game birds sitting in the path of the mower – something that concerned Harry Ferguson – a game flusher was offered. Fitted to brackets attached to the front axle was a length of angle iron, from which dangled nine separate long lengths of chain that dragged in the grass a tractor's length in front of the mower. The idea was that the birds would be put to flight before the mower reached them. Not many examples of the game flusher were sold and today it remains one of the rarest of Ferguson implements.

The side delivery rake was for getting two rows of hay into one. It was pulled at an angle, and six bars, each with 18 tines, were attached in a circle to a pair of PYO-driven rotating units, the action driving the bars and tines downwards and sideways to form the raking action.

ABOVE: A rare implement – the game flusher.

Somewhat old-fashioned in concept – causing one to wonder why it was offered as a Ferguson implement – was the hay sweep, which fitted on brackets attached to the front axle. Eight long tines were lowered to the ground and driven through the loose hay, gathering it up until a sufficient amount was collected, whereby it would be taken to an elevator that transported it to the top of the hayrick. Hay was removed from the sweep simply by reversing the tractor. This is another rare Ferguson implement of which few survive. Even more obscure and also somewhat of a throwback was the rick lifter, used to transport larger amounts of hay gathered into heaps. Operated from the rear linkage, this implement is so rare that apparently no pictures are available of it, nor any sightings.

Produced for overseas markets was the hay rake, which also harked back to earlier farming practice. The Ferguson version was mounted on the three-point linkage rather than being trailed. Thirty-two slender spring tines, plus two outer ones, gathered up the hay, with two skids preventing the tines digging into the ground. Using the hydraulic system to raise the rake released the hay. If the cut grass was to be used for making silage instead of hay, then buckrakes with either 10 or 12 tines were available, the tines being long tubular spikes with pointed ends. The buckrake fitted onto the three-point linkage using a different top link, which allowed tipping. Loading the buckrake was achieved by reversing into the cut grass with it lowered. When full it was raised and the grass was transported to the silage pit or heap, which was usually in the vicinity of the farmyard. Introduced later, the forage harvester eliminated the need for a buckrake as it collected and chopped grass directly from the cut row and fed it into a high-sided trailer, which enabled much more grass to be transported than could be carried on a buckrake.

When not concerned with crops or livestock, there was always something else to keep farmers occupied and one of these was maintaining fences, particularly for those that kept animals. Fences needed posts and so did gates – much larger ones. Holes needed to be

13. THE FERGUSON SYSTEM

LEFT: Before the advent of big balers, silage was taken from field to pit using a buckrake.

BELOW: The baler attached alongside the tractor.

opened up in the ground, so to make the task much easier Ferguson produced a post hole borer or digger, as it was termed. This attached to the three-point linkage, with the dedicated split top link attached to a cross member fitted between the two lower arms and finishing either side of the digger gearbox, from which was suspended an auger – available in four different diameters, the design of the linkage enabled it to maintain a vertical position when lowered into the earth. Drive was taken from the tractor PTO.

There were plenty of options for other farm work involving the removal and transporting of soil. The earth mover featured a curved steel blade, similar to a bull-dozer blade, which fitted onto brackets beneath the front of the engine. Rods running alongside the tractor and

RIGHT: A special linkage ensured the auger of the post hole borer remained in a vertical position.

terminating on additional brackets attached to the lower link arms allowed for raising and lowering of the blade hydraulically. A useful attachment for light levelling and trench back filling, the blade could be used in line with the front of the tractor or angled left or right, depending on requirements.

A similar device was the dual-named earth leveller and blade terrace, which fitted onto the rear linkage and could also be angled up to 45 degrees both left and right. Carrying out similar but more precise functions was the grader blade, which also fitted onto the three-point linkage but sat to the rear of the tractor. It too could be angled to the left or right, and also reversed for soil pushing. To ensure accurate soil levelling, it was necessary to add an optional wheel, fitted to the rear of the main beam.

ABOVE: The earth mover blade could be angled to the right or left.

13. THE FERGUSON SYSTEM

To remove and transport quantities of soil, the earth skip was available. A simple three-sided steel box, the open side formed an edge that skimmed the earth and it fitted onto the three-point linkage without the need for additional parts. In the raised position and full of soil, it was emptied by means of a lever-operated trip mechanism which allowed it to tip forward.

Useful for transporting concrete and other loose materials was the dump skip. Looking very similar to a dumper skip and with a 10 cwt capacity, it sat within its own framework attached to the three-point linkage. A trip bar allowed it to tip for emptying, with the skip restrained by chains when fully tipped.

Another rear-mounted implement with a more industrial nature was the crane, which featured an adjustable jib constructed of two steel box sections, one able to slide within the other and secured in one of the five-inch steps by pins, with a steel hook on the end for attaching the load to. Welded to the larger box section was a tubular steel framework that attached to the two lower links. At its shortest jib length of 4ft 2in, the crane had a capacity of 650lb and at its longest jib length of 6ft 11in one of 350lb. Raising and lowering of the jib was by means of the normal hydraulics. The crane is another of the rarer Ferguson implements and again commands a high price when up for sale.

The implements sold by Ferguson were all made by other companies, many of which specialised in specific products, although from looking at the implement its origin would be unknown. Two implements in particular went against this protocol, in that the main part of the implement consisted of a branded item. The Hydrovane air compressor was bolted onto a simple tubular steel frame that fitted onto the hydraulic linkage. Drive was taken by flat belt from a drive belt pulley fitted to the PTO. The intended main use of the compressor was to power reciprocating blade hedge cutters, but many other air-powered tools could be also used. The Berisford water pump was also fitted into a tubular steel frame, again driven by flat belt from the belt pulley. Both of these items are collectable and achieve good prices.

The Hesford winch was used mainly for working with felled trees by removing them from difficult locations. It fitted onto the three-point linkage and driven from the PTO. Levers controlling the reel clutch and a brake could be operated from the driver's seat, while a substantial ground anchor prevented the tractor being dragged backwards when the winch was in use.

For those burning wood for heating, the cordwood saw proved invaluable for cutting timber into logs. This was another implement

ABOVE: Dump skip was a useful implement for building maintenance.

ABOVE: Perhaps a bit big for a small tractor, the Hesford winch.

driven by flat belt from the belt pulley and the three-point linkage for transport, sitting on its own frame when lowered into position. Timber was placed onto a pivoting table which could be moved towards the circular saw blade against the resistance of springs, and the timber moved along the table after each cut. The advantage of a tractor-mounted and driven saw was that it could be taken to the site of fallen trees to cut them up, rather than the laborious process of moving the tree.

Cleverly using the tractor's hydraulic system as a source of power, the jack raised the tractor, enabling wheels to be removed either to repair a puncture or to adjust the width. Constructed of tubular steel, the rear unit fitted onto the lower hydraulic arms and was linked to the front unit by chain. Raising the linkage moved the tractor forward and lifted front and rear simultaneously. Several slightly different versions were offered over the years.

Carried over from 9N accessories was the tyre inflation set, which used the compression of one cylinder to inflate tyres via a special adaptor fitted into a plughole and a length of hose with the tyre valve connector on the other end. This could only be used with spark ignition engines. ●

ABOVE: Cordwood saw table was pushed forward against the resistance of two springs.

13. THE FERGUSON SYSTEM

ABOVE: Although several prototypes were built and tested, the combine never entered production.

ABOVE: One of the prototype combines at work.

ABOVE: Ferguson patents were listed on both tractors and implements.

ABOVE: Single furrow reversible, or butterfly plough as some term it.

Chapter 14
Other Implements and Accessories

As with the tractors, Ferguson didn't produce any implements in the UK but relied on outside specialist companies for that task. They were labelled as a Ferguson product when marketed and were the official Ferguson implements and accessories. As the popularity of the Ferguson tractor increased, a number of manufacturers saw an opportunity and introduced products which could be fitted to it, and these ranged from the quite small to the large.

When parking a Ferguson tractor, the brake pedal is depressed and held down to prevent movement by the action of a pawl engaging in a ratchet. Because of the proximity of the brake pedal and footrest, it is possible to knock the pawl out of engagement when dismounting the tractor. At least two companies, Adroit and Bulldog, made a product to overcome this and on both examples the pawl was lever engaged from a housing fitted on top of the gearbox casing, and once engaged could not be accidentally knocked out. Replicas of the Adroit device are still manufactured.

Collecting bales of hay or straw is a labour-intensive process and many companies in the late 1940s and 1950s introduced mechanical devices to speed up the process. One such product was the 'Manless', produced to handle the small rectangular bales which were the standard size at the time. Of tubular construction, this device fitted along the left side of the tractor and used just the left hydraulic arm to move it up and down.

Used mainly for manure handling, the Cameron-Gardner loader fitted at the rear of the tractor on to the hydraulic linkage, the fork being reversed into the manure heap before being raised for transport.

ABOVE: Taking the hard work out of lifting bales, the Manless bale handler.

ABOVE: Useful for manure shifting was this Cameron-Gardner rear loader.

BELOW: More of a conversion than an implement, the Twose roller.

Row crop work often requires the tractor to be driven at a slow speed; one that cannot be provided by using first gear on the standard gearbox. To overcome this, the Howard Rotovator offered a reduction gearbox that could be fitted into the existing gearbox casing and used the hole previously occupied by an inspection cover to fit the operating lever. Ferguson offered a more sophisticated reduction gearbox in a separate casing that fitted between the gearbox and axle casings.

Of a size that could hardly be called an accessory or even an implement, the Twose Tractamount Roller – available in both 3.5 and 4.5-tonne capacities for use on sports grounds or for compacting road materials – required the tractor to be driven onto it to provide the power source and driving control. Power to the rear roll was taken from an adaptor fitted to the rear wheels and the conversion was described as taking less than 10 minutes. It is doubtful if many of these rollers were sold but at least two survive in the hands of enthusiasts.

14. OTHER IMPLEMENTS AND ACCESSORIES

More of an accessory than an implement, the Bombardier half-tracks had little use in the UK but proved worthwhile to give extra grip in countries where the ground was snow-covered during much of the winter, and came to prominence during the trans-Antarctic expedition of the late 1950s. Two endless rubber belts were joined together at intervals of about three inches by pressed steel cleats riveted to each belt. The tracks ran over each rear tyre to a jockey wheel attached to a mid-mounted sub-frame. Steering was by a combination of the front wheels and use of individual wheel brakes. Another version of the tracks continued over larger front wheels. These tracks are another highly sought-after collectors' item.

Farming, by its very nature, exposes workers to the elements. While those doing general

LEFT: More effort saved – a sack lifter of unknown make fitted to a narrow tractor.

BELOW: The Cult-Harrow was produced by Horstman of Bath.

ABOVE: Used in Antarctica, half-tracks are a sought-after accessory.

work can take some sort of shelter from the very worst of the weather, tractor drivers have little option but to stick to the task, particularly when ploughing. With very few exceptions, no sort of weather protection was provided on tractors but in the 1940s this began to change. Several companies offered tractor cabs for a range of tractors, but mostly the Ferguson as this was a popular make.

The cabs offered provided only rudimentary weather protection and came in a variety of designs and construction materials, canvas and aluminium at the fore. None were designed as safety cabs to prevent injury to the driver. It would be several years before Massey-Ferguson would introduce a cab, and also before safety cabs would become mandatory. Because of the sometimes-flimsy nature of the early cabs and the fact they were not useable under this new legislation, very few have survived – and those that have are sought-after. •

RIGHT: A contemporary cab advertisement.

14. OTHER IMPLEMENTS AND ACCESSORIES

ABOVE: Early cabs only provided weather protection for the driver.

Chapter 15
Bigger is Better?

There's no doubt that the Ferguson TE20, its later derivatives and the implements of the Ferguson System did a sterling job in improving farming efficiency after the Second World War. The tractor and its implements were ideally suited to the hundreds of small farms dotting the British Isles, where a 10-acre field was considered large. But while the TE20 could pull a two-furrow plough with ease on most soils, it could only cope with a three-furrow one on light soil. Ferguson salesmen were aware of this problem and requested the company produce a larger tractor, so as to compete with the Nuffield Universal and Fordson Major.

Ferguson agreed to build a prototype larger tractor, to be known as LTX (large tractor experimental), and the first unit was ready by mid-1949. Over time, a total of six LTX tractors were produced and a range of larger implements made to complement the tractors. Design-wise they were a larger version of the TE20. Out of the six experimental tractors produced, four had petrol engines and two were diesel. Extensive testing took place on heavy Warwickshire soils and those present stated the LTX considerably outperformed the new Fordson Major.

ABOVE: Final versions of LTX prototypes had similar bonnets to the FE 35.

15. BIGGER IS BETTER?

ABOVE: An LTX photographed alongside a competitor, the Fordson Major.

ABOVE: Ferguson LTX alongside an American Massey Harris, which was the inspiration for the Massey-Ferguson 65.

The takeover of Ferguson by Massey-Harris to create Massey-Harris-Ferguson (and later Massey-Ferguson) was to prove the beginning of the end for the LTX project. For the US market the option of a tricycle front axle was required, which could not be produced with the Ferguson axle, and the cost of producing a new engine was also a limiting factor. Inevitably the project was scrapped, and most prototypes cut up.

A new large tractor from Massey-Ferguson was produced using a Perkins 4.192 diesel engine, modified 35 transmission and tinwork manufactured on tooling from the discontinued US-built Massey-Harris 40. The new tractor would be known as the Massey-Ferguson 65.

There is an ironic end to the LTX story. The Hiatt family, who farmed at Ufton in Warwickshire, had been using an LTX for a number of years with the approval of Ferguson and in the early years any problems were fixed by the factory. On one fateful day in 1970, the clutch went on the tractor and a member of the family rang Massey-Ferguson to ask if they could mend it. At Coventry it was realised that this experimental tractor should have been destroyed years before, and a team was quickly despatched to retrieve this last LTX and remove all trace of it. •

ABOVE: What the LTX became, Massey-Ferguson 65.

Chapter 16
Merger

In the aftermath of the court case with Ford and with the tractor business making a loss in 1953, Harry Ferguson sought a discussion with James Duncan, the president of Massey-Harris. This was with a view to building combine harvesters, which Ferguson had increasingly become interested in. This initial proposal escalated first into an offer for the Ferguson manufacturing plant in the US and then to a complete merger with all of the Harry Ferguson businesses.

Massey-Harris, a Canadian manufacturer of farm equipment, was formed in 1891 from the merger of the Harris and Massey harvesting equipment businesses. The takeover of Ferguson was not to everyone's liking and, as in many cases of mergers, some at both Massey-Harris and Harry Ferguson Ltd didn't particularly agree with it. The merger wasn't exactly what Harry wanted either; he would have preferred his tractors to be built by a much larger company but apart from in the Ford years, none were particularly interested. Initially the new company was called Massey-Harris-Ferguson, later abbreviated to Massey-Ferguson.

Sir John Black, of the Standard Motor Company, was initially reluctant to deal with

ABOVE: The Black Tractor forms the centrepiece of the Massey-Ferguson stand at an agricultural show.

the new company and intended Standard to build its own tractor, to the extent that a prototype was built. But in an about-turn, a 12-year agreement was signed for Standard to build Massey-Ferguson tractors at Banner Lane.

The merger made Harry Ferguson a very wealthy man at 68 years of age. He stayed with the new company for only a year and then took no further part in the tractor business, instead concentrating on the development of road cars, with the adoption of four-wheel drive and anti-lock braking systems among other developments.

Through a new company, Harry Ferguson Research, two prototype cars were built and proved to be superior in certain aspects over cars built at the time by other manufacturers. Despite this early promise, the cars did not go into production, though some of the systems developed were incorporated into production cars and a steady business of converting vehicles to the four-wheel drive system by the company continued. ●

Chapter 17
Across the Ice

One of the more outstanding uses of the Ferguson tractor wasn't agriculture-related at all and didn't involve soil cultivation or crop production. While being confined to his sleeping bag in 1949 due to adverse weather conditions on the Antarctic continent, the explorer Sir Vivian Fuchs came to the conclusion that if man had been able to reach the South Pole and return along the same route, then it should be possible to transverse the continent via the South Pole.

It would, he realised, take a great deal of organising and money – somewhere in the region of £250,000, he estimated – and would not be possible with just dog sled teams. Mechanical transport and aerial reconnaissance would be required. To put things into perspective, the journey was equivalent to travelling from London to North Africa – about 2000 miles. As several territories would be crossed on such an expedition, he thought those countries should be invited to participate and the expedition became known as the Commonwealth Trans-Antarctic Expedition.

Planning and fundraising took several years, with early recruits office-bound on those duties. Finances were considerably helped by a donation of £185,000 from the British Government, and free fuel and lubricants donated by BP. The Queen agreed to become patron of the expedition, the purpose of which was to survey the continent and also gain scientific evidence.

In November 1955 Canadian sealer Theron sailed from London with the advance party.

ABOVE: Antarctic in miniature, a diorama replicating the crossing.

Arriving in early 1956 in the Weddell Sea, a base site was formed on the edge of the sea and named Shackleton, after the explorer of the same name. Two Ferguson tractors and a Weasel were unloaded from the ship. One of the Fergusons was equipped with a high-lift loader to facilitate loading stores, of which there were 300 tonnes to move to the base. Severe weather hampered operations. When the wind had died down, it was discovered that some of the sea ice had broken away and one Ferguson and 300 drums of fuel were lost. The remaining supplies had to be rationed to last the winter, although there was enough food to last three years. Theron returned to London.

In November 1956 the newly built Danish polar vessel Magga Dan set sail for Shackleton base with more team members and supplies. To map out and create two fuel dumps on a route to the Pole from the opposite side of the continent, a base was established on the edge of the Ross Sea and named after the explorer Scott. This expedition was led by Sir Edmund Hillary.

To aid the cost of the expedition, Massey-Ferguson and the New Zealand agents, C B Norwood Ltd., loaned five Ferguson tractors; two to be used on moving stores at the base. It was decided that all members of the Hillary team should be adept at tractor driving, and the area of the Tasman glacier of New Zealand was used as a training ground in July 1956. The party sailed on December 21,1956 on Endeavour to the Ross Sea, arriving in early January 1957. From the Scott base, a tractor route to the Pole was found and two fuel dumps established, using a Ferguson fitted with half-tracks for reconnaissance.

Unsurprisingly, it was found that the Ferguson in standard form was unable to get enough traction on the snow and ice. Half-tracks improved traction but the best results were obtained when full tracks were fitted. Both types of track were of the Bombardier type; endless rubber belts running over pneumatic tyres and fitted with steel cleats, larger front wheels fitted when full tracks were used.

ABOVE: All members of Sir Edmund Hillary's team had to be adept at tractor driving and practising took place in New Zealand.

The Ferguson tractors used were all petrol versions and averaged less than two miles per gallon of fuel. They were fitted with reduction gearboxes and the HT wiring was replaced with silicon leads to cope with the low temperatures. Cold weather batteries were fitted, together with stronger starter motors, and the brakes were waterproofed. Several modifications were made once the tractors were on Antarctica. Rollover bars were fitted, and canvas screening added to form rudimentary cabs to keep out some of the wind, radius arms were strengthened with angle iron, and additional material was welded to the track cleats.

Three tractors journeyed to the South Pole, roped together and pulling a total load of three-and-a-half tonnes. In the book 'The Crossing of Antarctica' the following is written about the Ferguson tractors: 'These lightweight tractors were not as fully prepared for low temperatures as the other vehicles, but they carried out a most noteworthy journey. Their decreased power and traction at altitude explains the somewhat low payload in relation to fuel consumption. The harder the snow surface the better the traction. Although more suitable and useful for base work, they could render excellent service for short-range independent polar journeys at reasonable altitudes. Their extreme reliability enabled the Ross Sea party to reach the Pole.'

At least two of the Polar Fergusons still exist. One is in a museum in New Zealand and one is in the Massey-Ferguson museum in France. •

ABOVE: Modified full-tracks proved to be the best at coping with the Antarctic conditions.

LEFT: Open top cabs were constructed on Antarctica to provide protection from the wind.

ABOVE: Stores were transported on sledges each carrying about 30 cwt.

ABOVE: Tractors equipped with high lift loaders proved invaluable for unloading stores at both bases.

17. ACROSS THE ICE

ABOVE: A restored Antarctic tractor sits in the Massey-Ferguson museum.

Chapter 18
Ferguson in Miniature

The young and old are fascinated by miniature versions of full-size tractors and vehicles, and a few manufacturers realised there would be a ready market for models of the Ferguson tractor.

One of the first was Airfix, who added it to their range of plastic model kits in 1:20 scale, with the separate parts needing to be glued together with polystyrene cement to build the tractor. Initially the parts could be any one of a variety of colours, as a lot of recycled material was used for the mouldings. The Airfix Ferguson stayed in production from 1949 until 1959 and was sold in a variety of packaging during that time. While fairly simple in construction, almost bordering on the crude, this kit is now highly sought-after and prized by collectors. Such is the appeal of the tractor that in 2015 another manufacturer produced a plastic model in kit form of 'the little grey Fergie', this time in the smaller 1:24 scale.

As to actual toys, there was a limited number available. Benbros produced a tractor with driver in about 1:43 scale that was simple in its construction, and Mettoy produced something similar. In the larger scale of 1:16 came a splendid clockwork-powered tractor from Birmingham toy manufacturer Chad Valley. Relatively few

ABOVE: A number of model and toy manufacturers have produced miniature versions of the Ferguson tractor over the years.

18. FERGUSON IN MINIATURE

ABOVE: Amazing detail in a scale model tractor with high lift loader and rear weight.

were made – possibly because of cost – and they are now also highly collectable. Surprising perhaps is the fact the TE20 was not produced by that great British institution Dinky Toys, even though they produced miniatures of the Massey Harris tractor and also the Field Marshall.

The Ferguson range of implements was extensive, but no toy manufacturer took on the challenge of making toy versions. There is a subtle difference between toys and models, the former designed to be played with and the latter to be admired. Brown's Models produced a model kit of the Ferguson A to a scale of 1:32 in the 1970s. In the late 1980s, ScaleDown Models began to produce highly detailed model kits, also in 1:32, of two versions of the Ferguson TE20 tractor, and the kits are still available today.

Despite being a pioneering tractor, few models of the Ferguson A have been produced. At the end of 2018, specialist model maker G and M Originals announced a limited edition of 100 examples of the Ferguson A on pneumatic tyres in 1:32 scale. They also announced limited editions of the 'Black Tractor' in both 1:16 and 1:32 scales. Even rarer is the unique large-scale model of the 'Black Tractor', commissioned by farmer and enthusiast Jim Russell.

With the advent of cheap-volume manufacturing in China, several companies saw this as an opportunity to produce well-detailed models in 1:32 and 1:16 scale at advantageous selling prices. The number of model implements produced though still falls below that of the full-sized examples, and in the main it is left to specialists to produce just a handful of products. ●

ABOVE: This large-scale Ferguson TE20 is a late-1940s toy rather than a model, still very collectable though.

ABOVE: Very few models of the Ferguson A have been produced. This limited edition scale model is by G & M Originals, of Nottingham. *G & M Originals.*

18. FERGUSON IN MINIATURE

ABOVE: G and M Originals, of Nottingham, made models of the 'Black Tractor' in a limited run. *G & M Originals.*

Chapter 19
The Legacy

The Harry Ferguson name became legendary long ago and today, although not topical, that name and its legacy live on.

While most modern tractors are huge powerful machines with four-wheel drives and poles apart from the A, 9N and TE20, more than 85 per cent of them now attach implements via a hydraulic three-point linkage – first introduced in the 1930s by Ferguson.

The TE20 and its successors were so suited to small farms that their use on a smallholding or so-called hobby farm is still very relevant today. The tractors and implements are plentiful and fairly inexpensive, and numerous companies supply parts to keep the tractors running in top condition.

Those who remember the tractors from their younger days can also indulge themselves by building up collections of tractors and implements and showing them at events dedicated to the preservation of vehicles. Clubs cater for Ferguson enthusiasts who, besides tractors and implements, often collect sales brochures and other Ferguson-related items.

One man who combined two aspects of Ferguson enthusiasm was Harold Beer, of north Devon. He used the Ferguson tractor his mother bought in the 1950s, together with a 35 also bought new by his family later, on his farm. Harold built up a considerable collection of Ferguson items which are now in the possession of his widow, Eileen, a fellow Ferguson enthusiast.

ABOVE: A number of companies produce parts to keep early tractors running.

19. THE LEGACY

ABOVE: Because of the small number produced, the Ferguson A remains a favourite among collectors.

ABOVE: Both of these tractors, decades apart, use Harry Ferguson's three-point linkage with hydraulic depth control.

ABOVE: One of two clubs for Ferguson enthusiasts is the Ferguson Club.

ABOVE: Friends of Ferguson Heritage caters for Massey Harris and Ferguson enthusiasts.

19. THE LEGACY

ABOVE: Two restored Ferguson Model As sit in an enthusiast's collection.

Harold's late cousin, Ernie Luxton, used to have a Ferguson tractor, which was also bought new by his family in the 1950s. Harold had so much equipment in a useable condition that the implements have been shown at work in a series of videos entitled 'Ferguson On The Farm', where small areas of a variety of crops were grown and worked specifically for the filming.

A near neighbour is Michael Thorne, who is equally passionate about Ferguson products, and assumed a different approach to collecting. Mike first drove a Ferguson tractor in his teens and was involved in farming in his earlier career. Many years later he bought a Ferguson TED 20 and became fascinated by it – so much so that he dispensed with other non-Ferguson tractors he had collected to concentrate on just the one make. From this small start, he has built up a large collection of some 50 tractors, together with implements and many associated items and, more importantly, a number of prototype tractors. At one stage he even seriously thought of going as far as reconstructing an example of the LTX. To display the collection to a high standard, Mike has constructed impressive buildings in which to house them and allows fellow enthusiasts to visit by appointment.

The 'Black Tractor' sits in the Science Museum in London. At the AGCO factory in France sits the first Model A, formerly owned by Thomas MacGregor Greer, and 'Sue', one of the tractors used on the trans Antarctic expedition.

On the Isle of Wight, descendants of Harry Ferguson have built a small museum. In Northern Ireland a blue plaque is displayed on the wall of Harry's birthplace, and not too far away is a memorial garden containing a specially commissioned sculpture of Harry leaning on a gate.

As splendid as this all is, for many the 'little grey Fergie' will always be a memorial to the determined inventor. ●

ABOVE: Mike Thorne's collection is housed in purpose-built buildings.

ABOVE: As expected, a large turnout for the last Banner Lane factory open day.

19. THE LEGACY

ABOVE: Crowds line the drive at Banner Lane open day (a crane is a sought-after implement).

ABOVE: Enthusiasts eye up two rare implements, the kale cutrake and cult-harrow.

ABOVE: A Ferguson combine gets paraded at the last Banner Lane factory open day.

ABOVE: Ferguson tractors are a familiar sight at ploughing matches.

19. THE LEGACY

ABOVE: This sculpture of four Ferguson tractors sits outside a tractor dealership on a road leading into Dublin.

Bibliography

Aircraft and Aerospace Manufacturing in Northern Ireland. Guy Warner & Ernie Cromie. Colourprint Books.

The Crossing of Antarctica. The Commonwealth Trans-Antarctic Expedition 1955-58. Sir Vivian Fuchs and Sir Edmund Hillary. Cassell.

The Ferguson Album. Allan T Condie. Allan T Condie Publications.

Ferguson Implements and Accessories. John Farnworth. Japonica Press.

Ferguson TE 20 in Detail. Mike Thorne. Herridge & Sons.

Ford, an unconventional biography of the two Henry Fords and their times. Booton Herndon. Cassell.

The Ford Dynasty. James Brough. Cassell.

Ford Farm Tractors. Randy Leffingwell. MBI Publishing Co.

Harry Ferguson. Norman Wymer. Phoenix House Ltd.

Harry Ferguson and Henry Ford. John B Rae. Ulster Historical Foundation.

Harry Ferguson. Inventor and Pioneer. Colin Fraser. John Murray (Publishers) Ltd.

Harvest Triumphant. The Story of Massey-Harris. Merrill Denison. The Falcon Press Ltd.

Pioneers and Inheritors: Top management in the Coventry motor industry 1896 – 1972. Steven Morewood. Centre for Business History, Coventry Polytechnic.

Vintage Tractor magazine. Tim Bolton. R T Bolton. Various issues.

Index

Adroit 86
Airfix 101
Allis Chalmers 17, 27, 30, 34
Antarctica 86, 96, 98, 108
Arps Corporation 61
Art Deco 30

Banner Lane, Coventry 37, 38, 95
Belfast Telegraph 8
Beer, Harold 105
Beet lifters 70
Benbros 101
Berisford water pump 82
Black Tractor, The 18, 20, 102, 108
Black, Sir John 36, 38, 94
Blackhawk 61
Bristol aeroplanes 37
British Board of Agriculture 13
Brown's Models 102
Bombardier tracks 88, 97
Bulldog 86

Cabs 89
Canada 7, 14, 94, 96
Cameron-Gardner loader 86
Chad Valley 101
Chambers, John 18
Clausen, Leon 27
Cockshutt plough 9
Continental Motors 39
Cordwood saw 82
Cork, Ireland 13
Court case, Ford 35, 94
Coventry Climax 20
Cripps, Stafford 37
Craven Wagon and Carriage Works 20

David Brown 18, 20, 26, 28
Dagenham, Essex 23, 24, 28
Dearborn, USA 14, 27, 34, 58
Detroit, USA 13-14, 38, 44
Dinky Toys 102
Duncan, James 94
Duplex plough 14, 27, 56

Earth leveller and blade terrace 81
Earth mover 80

FE 35 42, 105
Ferguson A 24, 39, 40, 56, 61, 62, 102
Ferguson A implements 56, 56, 62
Ferguson Brown Limited 20, 26
Ferguson, James and Mary 7
Ferguson, Joe 7, 9
Ferguson, Henry George (Harry) 7-9, 11-21, 26-29, 34-38, 44, 47, 50, 52, 55, 74, 78, 94-95, 105, 108
Ferguson, Maureen 15
Ferguson Sherman Incorporated 15, 17, 27
Ferguson System 31, 35, 55, 58, 61, 66, 72, 78, 91
Fertiliser 68-69, 72
First World War 9, 11
Ford 13-15, 17, 24, 27-30, 34-35, 37-40, 44-45, 48, 58, 61, 94
Ford 2N 34, 37, 38
Ford 8N 34-35, 44-45
Ford 9N 30, 34, 36-40, 45, 56, 58, 61-62, 105
Ford 9N implements 58, 61
Ford, Edsel 34,

Ford, Henry 11, 13-14, 18, 27, 29, 34
Ford, Henry (II) 34, 35
Ford Motor Co. 34-35
Ford V8 27, 30
Fordson E27N 48
Fordson F 13, 30, 56
Fordson Major 91, 93
Fordson N 23, 30, 34
Fordson tractors 13-18, 21, 23, 30, 34, 48, 56, 91, 93
France 45-46, 98, 108
Freeman-Saunders 42, 52
Fuchs, Sir Vivian 96

G and M Originals 102
Game flusher 78
General Motors 17
Government (UK) 11, 13, 37, 96
Greer, Archie 14, 15, 17
Greer, Thomas MacGregor 7, 9, 13, 19, 24, 108

Handshake Agreement, The 28, 34
Harrows 62, 65, 72
Harry Ferguson Ltd 9, 13, 94
Harry Ferguson Research 95
Hay lades 76
Hay sweep 79
Hercules engine 18, 20
Hesford winch 82
Hillary, Sir Edmund 97
Hoes 66, 72
Horstman of Bath 65
Howard Rotovator 87
Hydrovane air compressor 82

International Harvester 30
Irish Board of Agriculture 11
Isle of Wight 108

Jeep 50
John Deere 30

Kale 70-71
Ki-Gas 42

Liney, Nigel 52
Loaders 74
Loddon Engineering 47
London, UK 13, 37, 96-97, 108
LTX 91, 93, 108
Luxton, Ernie 108

Manless, The 86
Manure spreaders 72
Massey-Ferguson 29, 89, 93-95, 98
Massey Harris 38, 74, 102
Massey-Harris-Ferguson 38, 94
May Street Motors 9
Meadows 52-54
Mettoy 101
Ministry of Munitions 11
Model A (Ferguson Brown) 20-21, 23, 27-30, 36, 56, 62, 65, 108
Model T Eros (Ford) 12, 13, 14
Morris, William 18, 36
Moto Tug 61
Mower 78

New Zealand 97-98
Northern Ireland 7-8, 14, 15, 16, 19, 108

Nuffield 36, 91

Overtime tractor 9, 11

Patents 14-15, 17, 26-29, 35, 37, 58
Perkins 50, 52, 93
Post hole borer 80
Potatoes 19, 67-70
PTO 61, 65, 69, 71, 78, 80, 82

Rakes 78, 79
Reekie, Gavin 50
Roderick Lean 15
Row crop thinner 66
Royal Aero Club 8
Royal Agricultural Society of England 13
Russell, Jim 102

Sands, Willie 11-18
ScaleDown Models 102
Second World War 34, 37, 91
Selene of Italy 48
Sherman, George 15
Sherman, Eber 15, 27
Skips 82
Skunk, John 15
Sorenson, Charles 13
South Pole 96
Spring tine cultivator 65
Standard Motor Co. 36, 37, 38, 52
Standard Vanguard 38-40, 52

TE 47
TE20 34, 39, 40, 42, 44-45, 50, 56, 61, 68, 91, 102, 105
TE20 implements 61
TE A20 40, 42
TE B20 40
TE C20 41
TE D20 41
TE F20 42
TE H20 42
TE J20 42
TE K20 41
TE M20 41
TE P20 41
TEA 47-48
TED 48, 108
Thorne, Michael 54, 108
Tiller 65
TO 20 44-45
TO 30 45
Trailers 74-76
Transport box 77-78
TVO 41-42, 47-48, 54
Two-row planter drill 69
Twose Tractamount Roller 87
Tyre inflation set 83

United States of America 7, 9, 11, 14, 15, 17, 18, 20, 28, 29, 30, 36, 44, 45, 52, 58, 69, 93, 94
Universal seed drill 68

Vapormatic 47
Vauxhall 9

Waterloo Boy 9
Weeding 66
Williams, John 15, 27-28
White, David 54